Lecture Notes in Computer S

Abdelkader Hameurlain
Farookh Khadeer Hussain
Franck Morvan
A Min Tjoa (Eds.)

Data Management in Cloud, Grid and P2P Systems

5th International Conference, Globe 2012
Vienna, Austria, September 5-6, 2012
Proceedings

Volume Editors

Abdelkader Hameurlain
Paul Sabatier University, Institut de Recherche en Informatique de Toulouse (IRIT)
118 route de Narbonne, 31062 Toulouse Cedex 9, France
E-mail: hameur@irit.fr

Farookh Khadeer Hussain
University of Technology, Faculty of Engineering and Information Technology
Centre for Quantum Computation and Intelligent Systems (Core Member)
Decision Support and e-Service Intelligence Lab (Core Member), School of Software
Sydney, Ultimo, NSW 2007, Australia
E-mail: farookh.hussain@uts.edu.au

Franck Morvan
Paul Sabatier University, Institut de Recherche en Informatique de Toulouse (IRIT)
118 route de Narbonne, 31062 Toulouse Cedex 9, France
E-mail: franck.morvan@irit.fr

A Min Tjoa
Vienna University of Technology, Institute of Software Technology
Favoritenstraße 9-11/188, 1040 Wien, Austria
E-mail: amin@ifs.tuwien.ac.at

ISSN 0302-9743 e-ISSN 1611-3349
ISBN 978-3-642-32343-0 e-ISBN 978-3-642-32344-7
DOI 10.1007/978-3-642-32344-7
Springer Heidelberg Dordrecht London New York

Library of Congress Control Number: 2012943352

CR Subject Classification (1998): C.2.4, H.2.4-5, H.2.8, I.2.6, H.2, C.2, C.4, H.3.4, H.4, H.5

LNCS Sublibrary: SL 3 – Information Systems and Application, incl. Internet/Web and HCI

Typesetting: Camera-ready by author, data conversion by Scientific Publishing Services, Chennai, India

Printed on acid-free paper

Springer is part of Springer Science+Business Media (www.springer.com)

Preface

Globe is now an established conference on data management in cloud, grid, and peer-to-peer systems. These systems are characterized by high heterogeneity, high autonomy and dynamics of nodes, decentralization of control and large-scale distribution of resources. These principal characteristics bring new dimensions and difficult challenges to tackling data management problems. The still open challenges to data management in cloud, grid, and peer-to-peer systems are multiple, such as scalability, elasticity, consistency, security, and autonomic data management.

The 5th International Conference on Data Management in Cloud, Grid, and P2P Systems (Globe 2012) was held during September 5–6, 2012, in Vienna, Austria. The Globe Conference provides opportunities for academics and industry researchers to present and discuss the latest data management research and applications in cloud, grid and peer-to-peer systems.

Globe 2012 received 15 papers from 12 countries. The reviewing process led to the acceptance of nine papers for presentation at the conference and inclusion in this LNCS volume. Each paper was reviewed by at least two Program Committee members. The selected papers focus mainly on data management (e.g., data storage, transaction monitoring, security) and MapReduce applications in the cloud, data stream systems, and distributed data mining.

The conference would not have been possible without the support of the Program Committee members, external reviewers, members of the DEXA Conference Organizing Committee, and the authors. In particular, we would like to thank Gabriela Wagner and Roland Wagner (FAW, University of Linz) for their help in the realization of this conference.

June 2012

Abdelkader Hameurlain
Farookh Khadeer Hussain
Franck Morvan
A Min Tjoa

Organization

Conference Program Chairpersons

Abdelkader Hameurlain	IRIT, Paul Sabatier University, Toulouse, France
Franck Morvan	IRIT, Paul Sabatier University, Toulouse, France
A Min Tjoa	IFS, Vienna University of Technology, Austria

Publicity Chair

Farookh Hussain	University of Technology Sydney (UTS), Sydney, Australia

Program Committee

Philippe Balbiani	IRIT, Paul Sabatier University, Toulouse, France
Djamal Benslimane	LIRIS, Universty of Lyon, France
Lionel Brunie	LIRIS, INSA of Lyon, France
Elizabeth Chang	Digital Ecosystems & Business Intelligence Institute, Curtin University, Perth, Australia
Qiming Chen	HP Labs, Palo Alto, California, USA
Alfredo Cuzzocrea	ICAR-CNR, University of Calabria, Italy
Frédéric Cuppens	Telecom, Bretagne, France
Bruno Defude	Telecom INT, Evry, France
Kayhan Erciyes	Ege University, Izmir, Turkey
Shahram Ghandeharizadeh	University of Southern California, USA
Tasos Gounaris	Aristotle University of Thessaloniki, Greece
Farookh Khadeer Hussain	Digital Ecosystems & Business Intelligence Institute, Curtin University, Perth, Australia
Sergio Ilarri	University of Zaragoza, Spain
Ismail Khalil	Johannes Kepler University, Linz, Austria
Gildas Menier	LORIA, University of South Bretagne, France
Anirban Mondal	University of Delhi, India
Riad Mokadem	IRIT, Paul Sabatier University, Toulouse, France
Franck Morvan	IRIT, Paul Sabatier University, Toulouse, France

Faïza Najjar	National Computer Science School, Tunis, Tunisia
Kjetil Nørvåg	Norwegian University of Science and Technology, Trondheim, Norway
Jean-Marc Pierson	IRIT, Paul Sabatier University, Toulouse, France
Claudia Roncancio	LIG, Grenoble University, France
Florence Sedes	IRIT, Paul Sabatier University, Toulouse, France
Fabricio A.B. Silva	Army Technological Center, Rio de Janeiro, Brazil
Mário J.G. Silva	University of Lisbon, Portugal
Hela Skaf	LORIA, INRIA Nancy -Grand Est, Nancy University, France
David Taniar	Monash University, Melbourne, Australia
Farouk Toumani	LIMOS, Blaise Pascal University, France
Roland Wagner	FAW, University of Linz, Austria
Wolfram Wöß	FAW, University of Linz, Austria

External Reviewer

Yannick Chevalier	IRIT, Paul Sabatier University, Toulouse, France

Table of Contents

Data Management in the Cloud

Cloud MapReduce and Performance Evaluation

Data Stream Systems and Distributed Data Mining

TransElas: Elastic Transaction Monitoring for Web2.0 Applications

Ibrahima Gueye[1], Idrissa Sarr[1], and Hubert Naacke[2]

[1] University Cheikh Anta Diop - LID Laboratory, Dakar, Senegal
{ibrahima82.gueye,idrissa.sarr}@ucad.edu.sn
[2] UPMC Sorbonne Universités - LIP6 Laboratory, Paris, France
hubert.naacke@lip6.fr

Abstract. Web 2.0 applications as social networking websites deal with a dynamic and various transaction workload. A middleware approach can be considered as a suitable solution for facing those various workloads. However, even if the middleware resources may be distributed for scalability and availability, they can be a bottleneck or underused when the workload varies permanently. We propose a solution that allows to add and remove dynamically resources of a distributed middleware. The proposed solution permits to handle transactions rapidly while using few middleware resources to reduce financial costs. Actually, we design an elasticity mechanism that distributes almost uniformly the transaction workload among the existing resources and adjusts the optimal number of nodes according to the workload variation. A simulation with cloudSim shows the effectiveness of our solution and its benefits.

Keywords: Elasticity, Transaction, Load balancing, Cloud computing.

1 Introduction

Web 2.0 applications are becoming more and more popular in the realm of computer science, economic, politic, and so forth because of their social and collaborative characteristics. Such applications are characterized by (1) a read-only intensive workload, (2) a number of users very important (tens of thousands), (3) a various kind of interactive applications such as virtual game, chatting and so on. These characteristics require to redefine the performance objectives of carrying on transactions. Actually, users create and manage their profiles by adding, day in day out, an increasing amount of data and new applications that leads to a various workload over the time. It is well known that managing such a complex workload with traditional OLTP mechanisms is not a viable and scalable solution. One of the characteristics of such kind of workload is that the number of users can be very high in a given period and fall down after a brief while. Therefore, using thousands of servers at any time is expensive and not efficient since most of the servers will be unused for a long period. However, using a very small number of servers has the drawback to make the system overloaded whenever the number of connected users is important. With this in mind, the

A. Hameurlain et al. (Eds.): Globe 2012, LNCS 7450, pp. 1–12, 2012.

best way to manage efficiently this frequent workload variation is to be able to increase/decrease progressively the resources allocated for handling transactions in such a way that the response time remains low or acceptable even if the number of users reaches millions of users. This property of a system or algorithm to run on a dynamic environment (variable amount of resources) is called elasticity and is one of the basic concepts in the cloud computing paradigm. In fact, cloud computing providers offer the possibility to add or remove resources based on the specific requirements of an application or a customer. However, cloud computing providers charge a customer for every resources used [9], which introduces new challenges for devising applications : actually, it is not only enough to process transactions as fast as possible but it is also strongly worthwhile to minimize the amount of resources for reducing the user fee.

Several studies have been proposed to deal with the elasticity [5,4,1,3,10]. Most of these solutions focus on the live migration of services that consists of adding new instances of nodes and thus, stop and move services from an overloaded instance to a new one. This migration mechanism is particularly suited for multi-tenant database systems and has unfortunately a significant additional cost. Precisely, this approach interrupts running transactions or services on a overloaded machine and shifts them to another machine, which may increase the response time and leads to a higher rate of transaction aborts. To avoid this problem, we propose a transaction management model that uses elasticity concept without data migration or tasks. To this end, we monitor the transaction routing process in such a way we execute entirely an incoming transaction on one node. Moreover, according to our best knowledge, existing solutions focus on the elasticity of the database layer (migrating the database partitions for load balancing, fault tolerance or auto-sharding). In contrast, we decide to focus only on the elasticity of the middleware layer since it faces the workload fluctuation before the database. It is worth noting that once the middleware is elastic, then it will be more trivial to handle the database layer elasticity (we leave this for future work). Moreover, live migration might be sensitive to the change of execution context. For instance, a transaction that reads the current system time before and after the migration may read inconsistent time values. Our solution does not suffer from this drawback since we avoid any transaction migration.

Furthermore, our solution relies on a previous work, namely, Transpeer [15] that is a distributed transaction middleware. Actually, the Transpeer middleware contains two kind of resources: 1) transaction managers that route queries, and 2) shared directory that store the metadata. The major drawback of this solution is the contention observed at the middleware under a heavy workload. This contention is noticed on the transaction manager level as well as on the metadata level. The main objective of our work is to avoid such a contention by managing the resources of the middleware in a elastic manner. In other words, we study in detail the sources of contention in order to know the type of resources to add for maintaining an acceptable response time. Our solution checks periodically the overall system status and adapts resources with respect to the workload size by balancing the workload and/or adding new resources.

The rest of this paper is structured as follows: Section 2 presents the architecture. Section 3 describes the middleware elasticity. Section 4 presents the validation of our approach while Section 5 highlights some work connected to ours and we conclude in Section 6.

2 Architecture

In this section, we describe the architecture of our solution that is designed to be highly available and to deal with a large scale distributed database. To meet these requirements, we distribute and replicate the components of the architecture in such a way that each component of the middleware is composed by a set (potentially high) of nodes. We have five components as shown in Figure 1, and each of them plays a specific role.

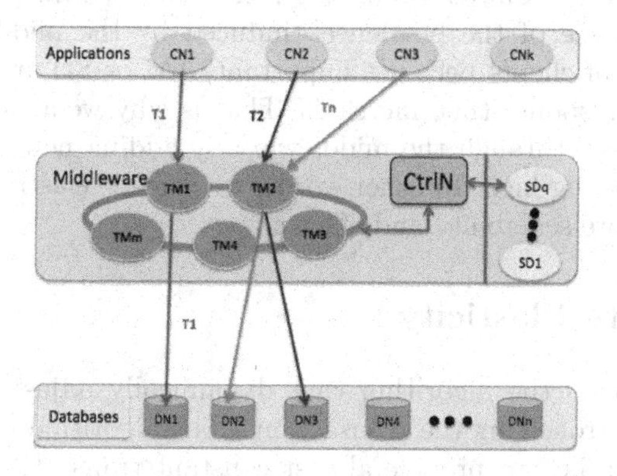

Fig. 1. System Architecture

- Application or client nodes (CN): the application is the component that produces the workload. Applications send requests that will be routed to the transaction managers.
- Transaction Managers (TM): It is the transactional middleware and the routing system core. It receives transaction requests and send them to the database system in a efficient way. Transactions are scheduled by the middleware in order to insure the consistency of the underlying database.
- Shared directory nodes (SD). It contains required informations (metadata) for routing efficiently the transactions, mainly the state of all system components.
- Database nodes (DN) : It is the component of the system that stores data. A database receives a transaction sent by the transaction manager and manages its execution, then sends results directly to the application that owns it. We mention that the database may have any kind of data structure, however it must has an API for receiving queries and sending results.

– Controller node (CtrlN). It monitors the overall system for detecting whether a TM or SD becomes a bottleneck or tends to be overloaded. The CtrlN node can use the watch mechanism provided by Zookeeper [7] to monitor the SD and TM loads.

TMs are gathered into a logical ring in order to facilitate their collaboration required to ensure consistency and to manage their failures. Consistency is ensured as in Transpeer [15] by maintaining a directed acyclic graph that allows update transactions to be executed at database nodes in compatible orders, thus producing mutually consistent states on all database replicas (see [15] for more details). Figure 1 depicts also how the five components of the system are coupled. The controller node, shared directory and transaction manager form the routing middleware. The other two components, application and database, located at the ends of the architecture, act as relay in regard to the external environment.

Furthermore, as we pointed out in the introduction, Transpeer solution did not scale out because of the bottleneck induced by the middleware. In fact, when the number of clients becomes important, TM or SD are overloaded and consequently, the response time increases. That is why we aim in this paper to avoid any bottleneck through the middleware by adding new TM or SD when the workload is getting high. In other words, we provide elasticity ability to the middleware for more scalability and efficiency.

3 Middleware Elasticity

This section presents the algorithm that dynamically adjusts the number of nodes involved in processing the transactions. The algorithm aims to make the middleware elastic, i.e., to provide almost constant transaction response time, whatever the workload that is submitted to the middleware. To this end, we design a solution for provisioning the middleware with machines (acting as TM or SD nodes) such that, there is enough nodes to handle the workload, but not too much nodes because we aim to minimize the overall number of nodes that we use. We do not need to add a new node if it is possible to balance the transaction workload among the existing nodes. In consequence, we first design a load balancing mechanism to distribute almost uniformly the load among the existing nodes. Then, we design a solution for elasticity that adjusts the optimal number of nodes according to the workload variation.

3.1 TM Load Balancing

Each TM is permanently waiting incoming transactions sent by several clients. We initially assign a TM to each client in a round robin manner in such a way that every TM manages the same number of clients. Since the clients send various transactions that may have different execution times, it is possible that a TM becomes overloaded compared to other TMs. Thus, to better balance the load, we adjust the number of clients associated with a TM, i.e., we move a client from an overloaded TM to a less loaded one. With this in mind, we proceed as follows:

- Each TM periodically informs the CtrlN about its current load status. The load value is a usage ratio (a value greater than one that indicates an overloaded node). Hence, the CtrlN gather the overall load status of the middleware and maintains the sliding average load L_i for each TM_i.
- We define two load values L_{down} and L_{up}. If L_i is in the range [L_{down}, L_{up}], then no load balancing process is made. This guarantees stability of the TM-to-client association, in case of small load fluctuations. The actual values of L_{down} and L_{up} are application specific. Intuitively, with a L_{up} value close to 100%, it may happen that a node is temporarily overloaded. On the other hand, a lower L_{up} value would prevent a node from being temporary overloaded, but in counterpart, it will cause the node to be globally under used.
- For each TM_i with a load $L_i > L_{up}$, we find the least loaded TM that can receive the extra load (i.e., L_i - L_{up}). For stability reason, we only consider candidate that has a small load ($L_{down} \leq L_i < L_{up}$). Moreover, if the amount of extra load can not be moved to one single TM, we split the extra load in several parts and assign each part to TMs in such a way that no TM candidate will become overloaded at the end of the load balancing process. If no candidate is found, then we add a new TM as described below.

3.2 Modifying the TM Load

The TM load is made of by transaction requests. Since, the transaction execution time is rather short, we do not attempt to move a running transaction by changing the TM which is responsible for it. However, we move forthcoming transactions by changing the TM that they will call later on. The main action to handle load balancing and elasticity consists of moving part of the load from a TM to another one. We define the primitive *MoveLoad(i, j, r)* that moves r% of the TM_i load to TM_j. In fact, CtrlN sends to TM_i a message ⟨move, r, TM_j⟩, where r is the load ratio to remove from TM_i, and has to be added on TM_j. Upon receiving this message, TM_i selects the k most recent clients (with k = n**number of active clients*), which represent the set of clients to move (denoted M). Subsequently, when TM_i sends a transaction result back to a client that belongs to M, it piggybacks a message that asks the client to use TM_j instead of TM_i, from now. This causes the client to close temporarily its connection with TM_i, and redirects its following requests to TM_j. Notice that the overhead in terms of communication is only one extra message from CtrlN to TM_i.

3.3 TM Elasticity

In order to provide elasticity of the TM layer, we investigate when to add/remove a node, and detail how to load a newly created node, or unload a node to be removed. We trigger the actions to add/remove a node, depending on the current average load L_i of node TM_i:

- If $L_i > L_{up}$ and no other TM can support the extra load, then we start a new node TM_j and move extra load from TM_i to TM_j.

- If $L_i < L_{down}$, and if there exists a set S of TM able to support L_i in such a way that each t in S has its load above L_i, and any t will not become overloaded afterwards, then we distribute the load among the TM in S. This sets to empty the node TM_i. Then, we notify the directory which initially assigns a TM to a new client, not to suggest TM_i anymore. Therefore, the idle TM_i node is shut down.
- If L_i is in $[L_{down}, L_{up}]$, then we do not change the number of nodes to avoid oscillations in the number of nodes.

3.4 SD Load Balancing and Elasticity

We remind that the SD nodes monitor the metadata access that a TM needs for routing a transaction. In case a SD node becomes overloaded, we cannot simply redirect the incoming requests to another less loaded SD node, because distinct metadata are attached to distinct SD node. We also have to transfer the part of the metadata from a SD node to another SD node. To this end, we rely on range partitioning mechanism to dynamically adjust the range of metadata that each SD is controlling. The n metadata are denoted by M_1, ..., M_n. Typically M_i is the metadata about the ith data element. Initially, we uniformly distribute the metadata among the m SD nodes, by partitioning the domain $[1, n]$ into m consecutive ranges. Let V denote a vector of dimension $(m + 1)$. Each SD_i is managing the metadata M_k such that $k \in [V_{i-1}, V_i[$. Then load balancing the SD follows a similar method as load balancing the TM. The CtlN periodically checks the SD load. Then for each overloaded SD_i, we proceed as follows for balancing the load or adding new SD:

- If (SD_{i-1}) is able to support the extra load (i.e., $i > 0$ and $L_{i-1} \geq L_{down}$), then the NCtrl sends a message to SD_i to move the extra load to SD_{i-1}.
- Else, if (SD_{i+1}) can support the extra load; (i.e., $i < n$ and $L_{i+1} \geq L_{down}$), then the NCtrl sends a message to SD_i to move the extra load to SD_{i+1}.
- Otherwise it creates a new SD node that will handle the extra load.

In the opposite, for each under-loaded node SD_i such that $L_i < L_{down}$, if there exists SD_j ($j = i - 1$ or $j = i + 1$) that can support the whole SD_i load without becoming overloaded, then we move SD_i load to SD_j. Therefore, we shutdown SD_i once all TM are aware of the change.

3.5 Modifying the SD Load

Since each SD controls a distinct range of the metadata, changing the metadata range that a SD controls will result in changing the SD load. Consequently, we define the *MoveSDLoad(i, j, r)* primitive that manages the load modification. Precisely, CtrlN sends to SD_i a message $\langle$move, r, $TM_j\rangle$ with r, the load ratio to move from SD_i to SD_j, and j is either $i-1$ or $i+1$. Upon receiving this message, SD_i update its vector V such that it releases the control on r% of the range it was controlling. Hence, SD_i moves the corresponding metadata to SD_j. Later on, the information about the metadata transferred from $SD_i$to SD_j will be propagated through the TM nodes on demand. When a TM ask SD_i for a metadata within the transferred range, the SD_i replies with a $\langle$redirect to $SD_j\rangle$ message that allows the TM to update its own V vector, then to directly connect to SD_j. We choose to propagate on demand rather than as soon as possible, in order to minimize the overhead of the primitive. Moreover, this primitive is general enough to handle load balancing among two consecutive existing nodes, as well as elasticity. Scaling up to more SD nodes, results in splitting the metadata range. Scaling down to less nodes results in merging consecutive ranges.

4 Validation

In this, section we validate our approach through simulation by using CloudSim [2]. CloudSim is a cloud computing environment that permits to get an infinite resources and the overall features we can find generally in a real cloud infrastructure such as data centers, virtual machines (VMs) and resource provisioning policies. It is an extensible simulation toolkit developed with JAVA language that enables both modeling and simulating a cloud computing systems. Therefore, we have extended/modified CloudSim classes in order to implement our approach. Since TransElas is a solution that adds elasticity capability in a previous work devised for handling transaction web2.0 (TransPeer [15]), our first goal is to highlight the impact of such an elasticity characteristic on the response time. To this end, we first compare TransElas latency versus the one of TransPeer. We focus on the latency of the TM nodes because they play a primary role in the transaction routing process. Then, we measure the elasticity of our solution and its ability to add a new TM only when needed. Finally, we measure the overall performance of TransElas middleware (both TM and SD) in terms of number of resources used.

4.1 Simulation Setup

The experiments were conducted on an Intel duo core with 1 GB of RAM and 3.2 GHz running under Windows 7. We had created via CloudSim a datacenter that contains 4 physical nodes. Each of them has the following characteristics: 2660 MIPS of CPU, 16 GB of RAM and 1 GB/s of bandwidth. On each physical node, we instantiate virtual machines (VM) as well as we need to face a various workload. In fact each TM or SD is associated to one VM, which corresponds to one instance of AMAZON EC2 (1600 MIPS of CPU and 1.7 GB of RAM). The workload is made of by a number of clients that varies during all our experiments. A client (NC) sends a transaction and waits until it gets the results before sending another transaction. All transactions access randomly data with the same granularity. This is possible because we assume that data are partitioned in such a way that a transaction processes only one single partition.

4.2 TransPeer vs. TransElas

As we pointed out before, we aim to compare the response time of routing an incoming transaction by using either TransElas or TransPeer. To this end, we start with 10 TM, and each TM can process simultaneously up to 10 transactions. Beyond this threshold, the TM is considered as overloaded and thus, a new TM is added. This threshold is chosen arbitrarily during our experiments but in a real life situation, it is fixed by the SLA (Service Level Agreement).

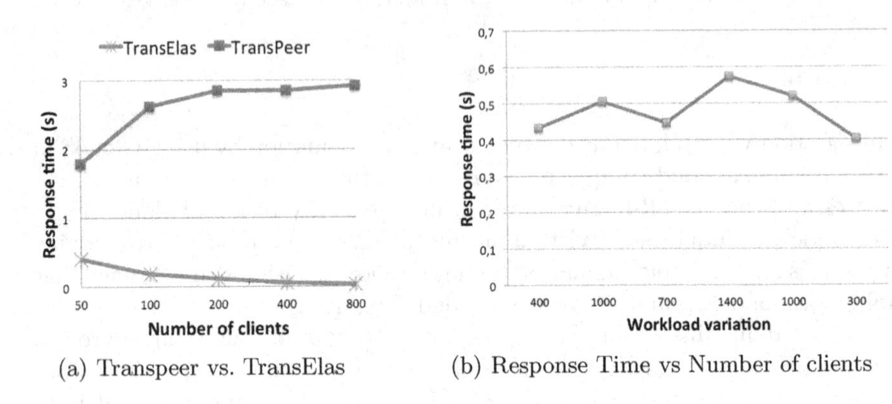

(a) Transpeer vs. TransElas (b) Response Time vs Number of clients

Fig. 2. TransElas overall performances

We vary the workload from 50 to 800 transactions and we report the transaction routing average response time on Figure 2(a). Results show that response time grows when the workload increases with TransPeer (red curve) while we observe the opposite with TransElas (blue curve). The reasons are that with TransPeer the number of TM are not increased when the workload becomes heavy. Therefore, some TM are overloaded and transactions are queued for a while before being routed. In the opposite, TransElas adds a new TM once a TM is overloaded, which gives more availability to route an incoming transaction as soon as it is received. This result shows all the benefits for adapting the number of resources *w.r.t* the workload.

4.3 TM Elasticity

In this experiment, we aim to evaluate the performances of the TM elasticity in terms of response time and number of TM used. First, we measure the response time when the workload (the number of clients) varies randomly from 300 to 1400, *i.e.*, the workload can increase at t and decrease at $t + \epsilon$ or vice versa. To guarantee an almost constant response time while minimizing the number of TM, we add new TM when some of them become overloaded, and we remove a TM when it is under-loaded. Figure 2(b) shows that the average response time varies slightly even if the workload varies randomly. For instance, with a variation of the workload from 700 to 1400 transactions, we notice an increase of just 0,1 seconds in the response time. This low increase is the direct result of our algorithm that add TMs progressively if the system tends to be overloaded. We observe also a decrease of the response time when the workload decreases. This is mainly due by the fact that we wait for a while before removing an underloaded node. Therefore, we maintain in the system for a while so many TM that gives more opportunities to process automatically an incoming transaction.

Second, we count the number of TM required *w.r.t.* workload variation. Our main goal in this experiment is just to measure the overall number of TM required for facing a growing number of clients. We consider 3 scenarios: *1) Scale-Up Only (SUO)* – TM are added when the workload increases; *2) Scale-Up and Scale-Down (SUSD)* – a new TM is added if a TM is overloaded as well as a TM is removed if it is under-loaded; *3) Elasticity and Load Balancing (ELB)* – we include a load balancing process: if a TM is overloaded, we try to balance its extra load first before trying to add a new TM.

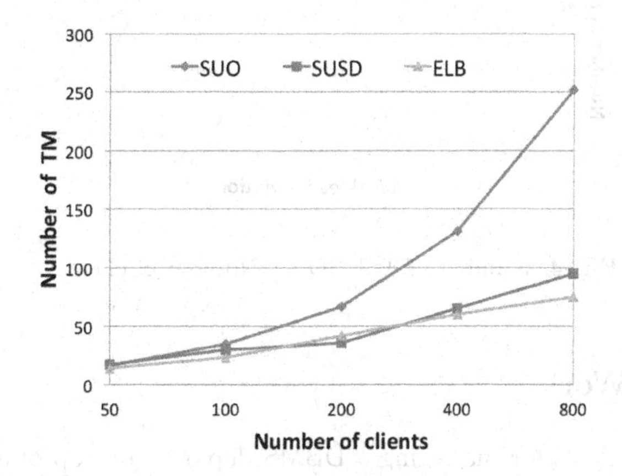

Fig. 3. Number of TM vs. Number of clients

Figure 3 shows that the ELB and SUSD require less TM than the SUO scenario. In fact, the ELB as well as SUSD scenarios add an remove TM *w.r.t.* to the workload variation. In the opposite, the SUO option do not remove TM even if the workload is low, thus, once a TM is added it lasts for all the experiment The ELB option works better when the workload becomes a bit more heavy (up to 500 clients) because it manages more efficiently the resources in such a way that a TM is added if there is no possibilities to balance an incoming transaction.

4.4 Overall Middleware Elasticity

In this section we aim to evaluate the performances of ensuring the middleware elasticity (both TM and SD). We use the ELB option described in the previous experiment since it requires less resources. Figures 4 depicts the overall number of resources required to face a varying workload (300 to 1400 clients). We recall that adding a new SD leads to split the amount of data it holds in two parts. When the number of clients varies, we observe that the number of resources used increases or decreases *w.r.t.* to the workload. The low number of resources used is mainly due by our load balancing mechanism applied both on TM and SD. In fact, when the number of clients decreases, our algorithm reduce the number of TM or SD while it increases the number of TM only if there is no possibilities to balance load of the overloaded TMs.

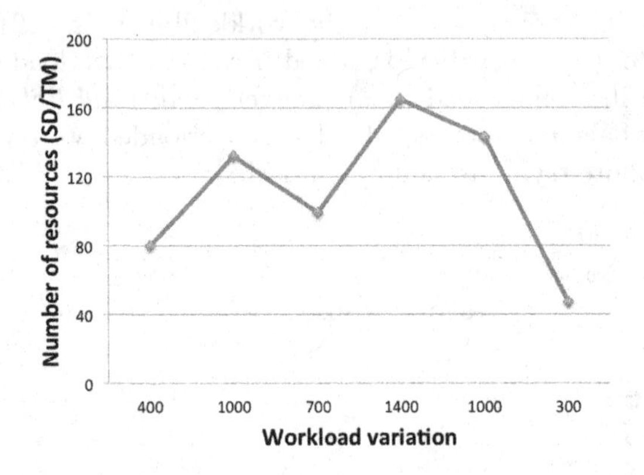

Fig. 4. Number of TM/SD vs. Number of clients

5 Related Work

One of the main goals for managing a DBMS deployed on top of a pay-per-use cloud infrastructure is to optimize the execution cost. Elasticity, *i.e.*, the ability to adjust and/or to reduce resources while the workload varies, is an essential issue to address for minimizing the cost. In this context, several studies have been proposed for dealing with elasticity [5,4,1,3,10]. Some of these works are based on the live migration approach [5,4,1]. Live migration consists to add new instances and then, to interrupt and to migrate services from the overloaded nodes to the new nodes. This migration approach is also used in the context of virtual machines and some studies has dealt with it [3,10]. However, the live migration of data and services from a node to another one may be critical : it requires to create new instances, to interrupt running services/transactions and to transfer them and part of some data. Beside the fact that these steps can increase the response time, they may induce to a higher rate of transaction aborts. In the opposite, our elasticity algorithm is processed for forthcoming transactions and not for running transactions, *i.e.*, we avoid to interrupt services from one place and to restart them on another. Moreover, solutions like [5,4] are black-box systems designed with a specific data model, namely a *key-value* model. Therefore, they do not act as a middleware that can be used with various kind of databases model as in the case with our solution.

Furthermore, elasticity is sometimes coupled with a load balancing process to minimize the number of resources used [5,4]. In addition, such a minimization of resources has an indirect gain on energy consumption and hence, reduces greenhouse gases emissions. That is why, some researches are conducted for developing new techniques of load balancing [14]. The objective of load balancing [13,11,12,8,17,6,16] is an efficient usage of resources by distributing the workload dynamically across all nodes in the system and thus, to avoid some bottlenecks or skews. In [12], the authors propose a new content aware load balancing policy named workload and client aware policy (WCAP). They use a unique and special property (USP) to characterize queries as well as computing nodes. The USP property helps the scheduler to decide the best suitable node for processing queries. This strategy is implemented in a decentralized manner and reduces the idle time of the computing nodes. The works done in [13] propose a load balancing policy for widely distributed web servers. The service response times is reduced by using a protocol that limits the redirection of queries to the closest remote servers without overloading them. Our load balancing algorithm acts almost as the cited works by tempting to reduce the response time, *i.e.* a load is transferred from A to B in order to minimize the waiting time on A if it is overloaded. However our approach aims to minimize resources while ensuring elasticity and it is based entirely on the overall middleware status: a resource is chosen to receive a load from another resource based on its local load and not to its proximity or a client property.

6 Conclusion

In this paper, we propose an elastic middleware solution for handling web 2.0 applications. Our solution monitors the resources usage within the middleware layer and add or removes resources based on the workload variation. The designed algorithms aim to make the middleware elastic, i.e., to provide almost constant transaction response time, whatever the workload that is submitted to the middleware. Actually, since the middleware layer is formed by two kind of components (TM and SD), we design a solution for provisioning the middleware with machines (acting as TM or SD nodes) such that, there is enough nodes to handle the workload, but not too much nodes because we aim to minimize the overall number of nodes that we use. That is why we devised also a load balancing mechanism to distribute almost uniformly the load among the existing nodes, and therefore, we reduce the number of resources. We validate our approach through simulation by using CloudSim and run a set of experiments that show promising performances in terms of resources used and response time. Outgoing works are conducted to evaluate our solution in a real cloud infrastructure.

References

1. Agrawal, D., El Abbadi, A., Das, S., Elmore, A.J.: Database Scalability, Elasticity, and Autonomy in the Cloud (Extended Abstract). In: Yu, J.X., Kim, M.H., Unland, R. (eds.) DASFAA 2011, Part I. LNCS, vol. 6587, pp. 2–15. Springer, Heidelberg (2011)
2. Calheiros, R.N., Ranjan, R., Beloglazov, A., Rose, C.A.F.D., Buyya, R.: Cloudsim: a toolkit for modeling and simulation of cloud computing environments and evaluation of resource provisioning algorithms. Software: Practice and Experience 41(1), 23–50 (2011)
3. Clark, C., Fraser, K., Hand, S., Hansen, J., Jul, E., Limpach, C., Pratt, I., Warfield, A.: Live migration of virtual machines. In: NSDI, pp. 273–286 (2005)
4. Das, S., Agrawal, D., Abbadi, A.E.: Elastras: An elastic transactional data store in the cloud. CoRR abs/1008.3751 (2010)
5. Elmore, A.J., Das, S., Agrawal, D., Abbadi, A.E.: Zephyr: Live migration in shared nothing database for elastic cloud platforms. In: SIGMOD (2011)
6. Gilly, K., Juiz, C., Puigjaner, R.: An up-to-date survey in web load balancing. In: World Wide Web, pp. 105–131 (2011)
7. Hunt, P., Konar, M., Junqueira, F.P., Reed, B.: Zookeeper: wait-free coordination for internet-scale systems. In: Proceedings of the 2010 USENIX Conference on USENIX Annual Technical Conference, USENIXATC 2010, p. 11 (2010)
8. Jasma, B., Nedunchezhian, R.: A hybrid policy for fault tolerant load balancing in grid computing environments. Journal of Network and Computer Applications, 412–422 (2012)
9. Kraska, T., Hentschel, M., Alonso, G., Kossmann, D.: Consistency rationing in the cloud: Pay only when it matters. PVLDB 21, 253–264 (2009)
10. Liu, H., Jin, H., Liao, X., Hu, L., Yu, C.: Live migration of virtual machine based on full system trace and replay. In: HPDC, pp. 101–110 (2009)
11. Lua, Y., Xiea, Q., Kliotb, G., Gellerb, A., Larusb, J.R., Greenber, A.: Join-idle-queue: A novel load balancing algorithm for dynamically scalable web services. An International Journal on Performance Evaluation (2011) (in press, accepted manuscript, available online)
12. Mehta, H., Kanungo, P., Chandwani, M.: Decentralized content aware load balancing algorithm for distributed computing environments. In: Proceedings of the International Conference Workshop on Emerging Trends in Technology (ICWET), pp. 370–375 (2011)
13. Nakai, A.M., Madeira, E., Buzato, L.E.: Load balancing for internet distributed services using limited redirection rates. In: 5th IEEE Latin-American Symposium on Dependable Computing (LADC), pp. 156–165 (2011)
14. Nidhi Jain Kansal, I.C.: Cloud load balancing techniques: A step towards green computing. IJCSI International Journal of Computer Science Issues 9(1) (January 2012)
15. Sarr, I., Naacke, H., Gançarski, S.: Transpeer: Adaptive distributed transaction monitoring for web2.0 applications. In: Dependable and Adaptive Distributed Systems Track of the ACM Symposium on Applied Computing, SAC DADS (2010)
16. Vo, H.T., Chen, C., Ooi, B.C.: Towards elastic transactional cloud storage with range query support. Proc. VLDB Endow., 506–514 (2010)
17. You, G., Hwang, S., Jain, N.: Scalable Load Balancing in Cluster Storage Systems. In: Kon, F., Kermarrec, A.-M. (eds.) Middleware 2011. LNCS, vol. 7049, pp. 101–122. Springer, Heidelberg (2011)

Recoverable Encryption through Noised Secret over a Large Cloud

Sushil Jajodia[1], Witold Litwin[2], and Thomas Schwarz[3]

[1] George Mason University, Fairfax, VA
jajodia@gmu.edu
[2] Université Paris Dauphine, Lamsade
witold.litwin@dauphine.fr
[3] Universidad Católica del Uruguay, Montevideo
tschwarz@ucu.edu.uy

Abstract. Encryption key safety is the Achilles' heel of cryptography. Backup copies with the escrow offset the risk of key loss, but increase the danger of disclosure. We propose *Recoverable Encryption* (RE) schemes which alleviate the dilemma. The backup is encrypted so that the recovery is possible in practice only over a large cloud. A 10K-node cloud may recover a key in at most 10 minutes, with the 5 minutes average. Same attempt at the escrow's site, a computer or perhaps a cluster, could require 70 days with the 35 days average. Large clouds are now affordable. Their illegal use is unlikely. We show feasibility of two schemes with their application potential.

Keywords: Cloud Computing, Recoverable Encryption, Key Escrow, Privacy.

1 Introduction

Users often want their data confidential. A popular tool is a high quality encryption. But users incur then the dangers of (encryption) key, hence of encrypted data, loss. Common cause is an IT infrastructure failure. Companies face also the loss because an employee departs with. Keys were lost in natural disasters, e.g. in the flooding of the basement of a large insurance (!) company. The keys & their backup, believed both safe there, became sadly unusable. Personal health data are more and more often encrypted by the owner. T he access to should remain however, even when the owner can't help. Encrypted family's data usually should survive the owner's sad disappearance with the keys, e.g., while boating in SF Bay. A key may alternatively be a trusted ID, e.g., a Diffie-Helman one. The loss of that one is also potentially catastrophic…

An approach to key safety is a remote backup copy with some *escrow* service [12]. It can be a dedicated company, an admin of the key owner company computer network, a cloud... Escrow simply retrieves the copy upon legitimate request. However, even a trusted escrow might reveal malicious or intruded by an adversary. A cloud can be unsafe against intruders. If an illegal key usurper discloses it, a big trouble may result. Consequently, users seem to refrain from escrow services. On the other end,

A. Hameurlain et al. (Eds.): Globe 2012, LNCS 7450, pp. 13–24, 2012.

there is the fear of the encrypted data loss. The dilemmas make many rather not using any encryption, [11]. This is also an invitation for trouble. By similar token, many users prefer a low quality authentication, e.g., a weak or repetitive password, despite the risk incurred…

The *recoverable encryption* (RE) in [12] was intended to alleviate this trouble. The backup is encrypted so that the decryption is possible without the owner, i.e., it supports the *brute-force* recovery, a *scalable distributed* one in fact for efficiency. The legitimate recovery is easy, while the illegal one is hard and dangerous for the intruder. In [12], RE was designed for the client-side encrypted data outsourcing to a large LH* cloud file. The backup was a shared secret, with shares spread randomly over many LH* buckets. The secret was then cumbersome at best to recollect illegally.

Here, we present more general schemes, collectively called RE through *noised secret*, RE_{NS} in short. A single computer or cloud node suffices as the backup storage. The client encrypts the backup so that the time the recovery could need at escrow's facility computer(s), appears long enough to deter illegal attempts. We usually note this time as D. The average recovery time is half of D as we'll show. The owner chooses D at will, but typically should plan months or years, depending on the trust to escrow and into the good fortune. The recovery request, whether from the owner or any other legitimate one, defines in contrast some maximal recovery time R, in seconds, minutes or, say, an hour. For the actual recovery, escrow must then invoke the service of a large cloud. An RE_{NS} scheme chooses the adequate cloud size, in number of nodes that we usually note N. N value is basically about linear with D/R.

In practice, N should be in thousands of nodes. Large clouds are now easily available. Google and Yahoo claim 100K-node clouds, Azur advertises millions of nodes… An unwanted recovery, by escrow or any other intruder must succeed with the illegal access to such an important resource. The cloud service providers already do their best against. Besides, an illegal use of such an "army" should leave marks in logs etc., creating the audit trail.

On the other hand, also intentionally, a legal request to a large cloud should be always somehow costly. The amount depends on D, R and actual cloud pricing. The requestor chooses then R according to the recovery urgency and acceptable cloud cost. Example we discuss later shows that RE over an 8K-node public cloud could currently cost a couple of hundreds of dollars on the average. By itself, it is then an obvious deterrent of unwelcome curiosity at escrow's side. For the recovery requestor, such cost should be acceptable as, to compare, a fraction of a usual car accident cost. It could further be subject to insurance. It could amount then to peanuts.

Technically, an RE_{NS} scheme hides the key within a *noised* secret. Like the classical (shared) secret, this one consists of at least two shares formed also "classically". At least one of these shares is further *noised*. It is then hidden within a large number M, called *noise space size*, of pseudo-random *noise* shares. The key owner actually sets up M as backup parameter, while D is at least linear with. With certainty in practice, only the noised share reveals the secret. The RE_{NS} recovery searches for the noised share through the noise ones by brute force. It may need to inspect all the shares and goes through at least half on the average. No brute force method known at present discloses a noised share in practice faster than RE_{NS} recovery does.

The search time on a single computer or cloud node, whether the worst or the average, scales up about linearly with M. To achieve the speed up over the cloud, an RE_{NS} scheme partitions the search. We propose two types of schemes, we call *static* or *scalable*. A static scheme fixes N upfront. A scalable one scales N incrementally, while partitioning. A static scheme may achieve the linear speed-up and, for every R requested, the smallest possible N. The constraint is the homogenous (same) throughput, i.e., the number of noise shares searched per time unit, at every node. A scalable scheme applies otherwise. It applies always in fact, but usually generates a somehow larger cloud for homogeneous nodes. It may still be preferred as more versatile and even cheaper money-wise.

The static scheme over 10K-node (homogenous) clouds may provide the speed up from, say, D of up to two months to R of ten minutes. A 100K-node cloud may scale down R further to at most 1m etc. The average timing decreases accordingly from $D/2$ to $R/2$. Here for instance, it could speed up from 1 month to about five min or, even, 30s. For the scalable scheme, the latter figure may be the same. However, usually both the worst and average times should be faster than R and $R/2$ respectively, while the cloud generated should be accordingly larger. For a homogeneous cloud, the differences should be typically about 30%. For a heterogeneous cloud, the time and cloud size should depend on nodes involved.

Below we analyze the feasibility of the RE_{NS} schemes. We define the schemes; discuss the correctness, safety and properties we have outlined. We present the related work. Space limitations force us to leave out many details in [14], fortunately only a click away for most readers today. No actual RE_{NS} scheme exists as yet. Our study shows nevertheless that these proposed fulfill the intent and should have numerous applications.

Section 2 presents the static partitioning and continues with the scalable one. Section 3 addresses the related work. In Section 4, we conclude and discuss the further work.

2 Recoverable Encryption through Noised Secret

The processing starts at the *client*, i.e., the key owner site. It produces the (noised) backup through *client-side encryption*. The client sends the backup to the (backup) *server* that is the escrow's computer. The client usually accompanies the backup with some access path to the actual data encrypted using the key. The *server-side decryption* recovers the key from the backup upon request.

2.1 Client-Side Encryption

The only secret data, say S, to back up, we consider below, unless we state otherwise, are a high quality encryption key. The highest quality keys for a symmetric encryption, the most frequent one in practice, are at present apparently those of AES standard, 256b wide. These keys are the default for what follow. Nevertheless, the client can backup shorter keys as well, e.g., DES ones or larger, e.g., 500b Diffie-Helman ones, often used also for identity certification, [13]. The crucial for safety is that *S* appears as a random integer.

To start with the backup, the client, say now *X*, chooses the dissuasive recovery time *D* that we already introduced. Technically, *D* is the owner's estimate of the longest time for an RE_{NS} recovery. This occurs on a single computer e.g., in the escrow's possession, or on 1-node cloud. The average 1-node recovery time should be then also dissuasive, being equal to *D*/2 at least, as it will appear. As we said, a wise choice of *D* should be in months at least.

While choosing *D*, *X* bears in mind that if the recovery occurs, the requestor fixes some maximal recovery time *R*, $R << D$, as we spoke about. The ratio *D/R* implies *N*. Basically, $N = D/R + \Delta(D/R)$. The actual *N* depends on accuracy of throughput prediction. We elaborate on this soon. Also, depending on the RE_{NS} scheme and cloud used Δ may be in practice negligible or up to about 30% of *D/R*, as we discuss in Section 2. Bottom line: *X* wisely chooses *D* large enough for the dissuasive purpose, but not larger, in the light of its interdependence with future *R*, the accuracy of the throughput estimate and the expected cloud bill if....

We now denote $w(y)$ the bit width (length) of integer *y*. After fixing *D*, *X* generates a random number, say s_0, where $w(s_0) = w(S)$. Below, we have $w = 256$ by default. Next, *X* calculates $s_1 = S$ XOR s_0. Afterwards, *X* applies to s_0 a good one way hash, say *H*, e.g., SHA 256 by default below. This produces the *hint*, say $h = H(s_0)$. By default, $w(h) = 256$ below, see [14] for more on hints.

X or the proxy algorithm, presumes further that a cloud node or the escrow's computer is a typical one according to the present technology, e.g., some 1-core Wintel node. *X* further knows somehow the throughput, we now denote as *T*, valuated per second, by default. It will appear that typically $T \cong 2^{20}$. *X* calculates a large integer $M = \text{Int}(DT)$. Next, *X* randomly chooses some integer *m* within interval $I = [0, M[$. After that *X* defines an integer *f* as $f = s_0 - m$. Then, *X* forms the so-called below *noised share* $s_0^n = (f, M, h)$. Finally, *X* forms the backup $S' = (s_0^n, s_1)$ and sends *S'* to escrow.

In essence, *X* first applies to S the "classical" secret-sharing, creating two (actual) shares s_0 and s_1. It is the common knowledge that $S = s_0$ XOR s_1. Then, *X* creates a finite *noise space* that is here interval *I*. Each value within *I* is a *noise*. The numbers $f, f+1 \ldots f+M\text{-}1$ become *noise* shares. *M* quantifies the number of noise shares and becomes noise space size, Figure 1. The actual share s_0 is one of the noise shares. This is easy to see for *I* above. One may identify s_0 from the noised one s_0^n through the successful match $H(f+k) = h$ for some "lucky" noise $k \in I$. The "lucky" *k* must be $k = m$. For any noise share $s \neq s_0$, hence $k \neq m$, the practical collision probability of a good one way hash is zero, hence $H(s) \neq h$.

Since *M* is finite, one may generate every noise in *I* and calculate the hash *H* of every noise share, e.g., by looping through $j = 0,1 \ldots M\text{-}1$. The lucky *k* must be encountered. Share s_0 will be found by the *m*-th match attempt $H(f+k)$?= *h*. The *m* value is however random and not in the backup. It may be equally likely anywhere in *I*. The brute-force recovery, not knowing *m*, has to guess it. Every noise in the noise space is equally likely, regardless of the way of picking up the noises. Any brute-force guessing may need to attempt then the match of even every of *M* noises and of

$M / 2$ at the average at least. Especially, since H is a one-way hash, so it is not possible at present to determine s_0 as $H^{1}(h)$. Also, since by general properties of the secret-sharing, a noise share other than the noised one cannot reveal the secret.

Practical values of M appear in hundreds of billions at least, e.g., $M = 2^{40} \div 2^{50}$, given the usual SHA 256 speed, [4]. These choices should let the 1-node recovery that is, to remind, a brute force one, to be typically by far too long in practice. As it will appear, any alternate brute force recovery methods leads in practice to at least as long worst-case and average timing. As we will elaborate, all this reasonably guarantees to the owner the correctness and safety of R choice. The former means that the 1-node brute-force RE_{NS} recovery eventually always returns s_0 and not faster than in the expected timing, provided the correct estimate of T. The latter means that no brute-force recovery could proceed faster.

By analogy to the secret-sharing terminology, we call the owner's action above *noised secret-sharing*. One can see that the characteristic difference is that the latter, unlike the former, generates one share with a noised part, i.e., s_0^{n} above. This part consists at least of the rightmost $w(M\text{-}1)$ bits of s_0. The actual value of the noised part is hidden somewhere within a purposely large noise space, I above. The value of a noised share, unlike that of a "classical" one is therefore purposely fuzzy. It is the noise width, itself an obvious function of M, that defines the fuzziness. As we stressed already, H is a one-way hash, not allowing at present to calculate s_0 as $H^{1}(h)$. As we also discussed, for any noise share s guessed as the (hidden) noised share s_0, known only by the hint h, the match attempt is the fastest practical way to find whether s is s_0.

Example 1. X wishes to backup some AES key S, with D set up to at least a month, i.e., 2^{22}s. The choice means that D appears to X long enough to deter an unwelcome recovery at escrow's site. It also means that X considers the eventual cloud if unthinkable happened, still feasibly large and reasonably priced. As it will appear, X may indeed expect $N \cong 8$ - 11K and the recovery bill of up to 400$ for usually practical enough $R \leq 10$m. Larger D would about proportionally increase the cloud size and likely the cost or forced longer R. Finally, the choice means that cloud size and the cost seem cumbersome enough to deter unwelcome cloud uses.

X also believes throughput T to be 2^{20} match attempts per second. One match attempt involves for a noise share s the comparison h ?= $H(s)$ and perhaps progress towards next noise if any. Given SHA 256 speed figures, such T seems reasonable for a popular, about 2 GHz, 1-core Wintel node, [4]. X fixes M to $M = 2^{42}$. This is the noise width w here and $I = [0, 2^{42}[$ is the noise space. Next, X produces a 256b long random integer as the actual s_0, calculates h and $s_1 = S$ XOR s_0 as the other actual share. Then, X chooses a random integer within $I = [0, 2^{42}[$, becoming m. Next, X computes f and forms the noised share s_0^{n}. Finally, X forms S' and sends it out.

The brute-force attacker of the backup can see in f the bits of s_0 which are perhaps out of the noise. These are at most all but the rightmost 42, i.e., the left 214b. The attacker may XOR those with same bits of s_1 recovering corresponding bits of S. For a high quality key, AES, especially, these bits do not let to infer the noised value of the rightmost 41 bits of s_1 and s_0. Equally, they do not make possible the determination of the rightmost 41 bits of S. The only use of the visible bits for the recovery is (i) to guess a noise, then, (ii) try out through a decryption attempt of the actual data whether S calculated that way is the right one. Any such verification we are aware of is by far longer than the hint match attempt.

The guessed value m may be indeed equally likely anywhere within I. There are then also "only" M equally likely noise shares. However, we hash over $w(s_0)$ being here 256b. By the property of a high quality one way hash, the value space of H is then very many times larger than M. There is no possibility at present to create a pre-computed inverted file that would have all or most of the hashes for a key. To the contrary, at the current stage of storage technology it could contain only a negligible in practice fraction of those. Same is true, more generally, for any other pre-computed decryption method known. Altogether, no guessing of the noised share can be at present faster than RE_{NS} match attempts, [14].

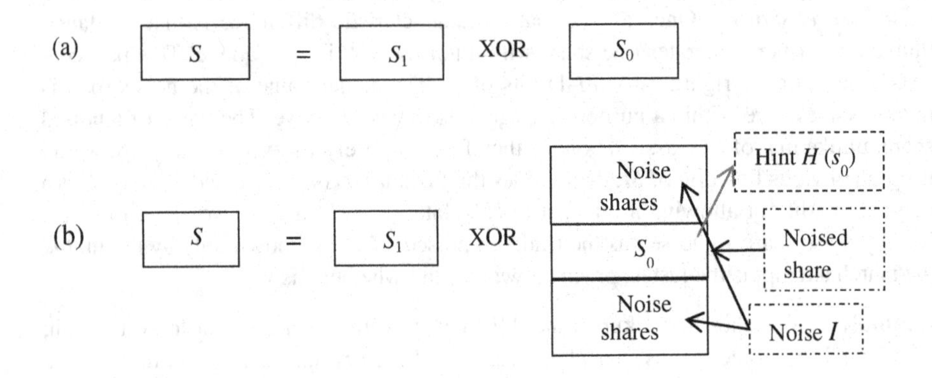

Fig. 1. 2-share (a) Secret-Sharing and (b) Noised Secret-Sharing

2.2 Server-Side Decryption

Overview

Upon receiving the backup data, the server, say E, simply stores them, together with some data that let to identify the legitimate requestor. Those data are outside our work. The backup requestor sends in particular R. As discussed, to determine S by brute force from noise shares one has to exhaustively search over I for a noise share s matching h what would mean in practice that $s = s_0$. This is the fastest way to find s_0. As we stressed, E may need to loop through all M-1 values, and should loop through $M/2$ on the average, since every noise in I is equally likely. This should be achieved for sure in at most R time units. E therefore partitions the match attempts over $N > 1$ nodes, with N chosen to meet this constraint. We now present two RE schemes we qualify respectively of *static* and *scalable*. The former chooses N upfront. The latter scales up N incrementally while partitioning. The static scheme works for a homogenous cloud i.e., with identical nodes and allocated to a single client at a time. Major public clouds can be of this type. The scalable scheme is designed mainly for a heterogeneous cloud. This can be a private or hybrid, or a public cloud, where several virtual machines may share a node. It applies to homogenous clouds as well, but typically should generate a larger cloud than the static scheme for the same *D* and *R*.

Static Partitioning

We call (*node*) load *L* the number of values among *M* assigned there for match attempts. As we already spoke about, we call (*node*) *throughput* *T* the number of attempts per time unit, a second typically. We call (*node*) *capacity* *B*, the size of the bucket of attempts that the node achieves in time *R*, i.e., $B = R*T$. Next, we call (*node*) *load factor* α the value L / B. The node performs for sure all its map attempts in time *R* only if $\alpha \leq 1$. The goal of the scheme is to get $\alpha = 1$ at every node participating in the calculus. This goal generates the smallest, hence in this sense optimal, cloud respecting R. The node *overflows* iff $\alpha > 1$. Our terminology is purposely that of scalable distributed data structures (SDDS) or files generally [15], to reuse the related properties.

When *E* requests the cloud service, *E* gets access to one node called *coordinator*, *C* in short. This starts the so-called *Init* phase of the scheme. *E* sends data $S" = (s_0{}^n, R)$ to *C*, keeping s_1 to avoid the disclosure of *S* by the cloud, for obvious reasons. *C* determines then its load factor, say $\alpha(M)$ induced by all *M* attempts. Basically, it knows somehow its *T*, computes trivially *B* and calculates $\alpha(M) = M / B$. If by any chances, it happens that $\alpha(M) < 1$, *C* performs the (entire) recovery. Otherwise, *C* calculates *N* as $N = \lceil \alpha(M) \rceil$ and requests the allocation of *N*-1 nodes. After that, *C* labels each node of the *N*-node cloud constituted in this way with the unique *node number i* ; $i = 0,1...$N-1. *C* itself becomes node 0.

Example 2. We continue with Example 1. As projected by *X* there, the requestor expects the time limit *R* on recovery to be 10m, say $R = 2^9 = 512$s. *C* knows *T* to be $T = 2^{20}$ per second. It computes *B* from *R* as $B = 2^{29}$ (match attempts per *R* in seconds). This leads to $\alpha(M) = 2^{42-29} = 8$K, far above 1. *C* requests therefore the allocation of 8K-1 cloud nodes. It labels itself node 0 and numbers the nodes 1,2,..8K-1.

Next, *C* advances to *Map* phase, named so according to the popular terminology. Here, *C* sends to every node *n* ; $n = 0...N$-1 ; *S"*, *n* value and program *P* defining *Reduce* phase. *C* also sends its physical, e.g., IP, address, so the node may eventually send back the positive match result. *P* requires each node *n* to attempt matches for every $k \in [0, M[$ such that $n = k \bmod N$. Node *n* basically loops then over the values $k = n, n + M, n + 2M...$ while $k < M$. If a match succeeds for some share $s = f + k$, the node sends *s* to *C* and terminates *P*. What it does anyway if no match attempt succeeded.

Some node must encounter *s*, [14]. Once *C* gets it, *C* starts *Termination* phase. It forwards *s* to *E* as s_0. Also, *C* requests the de-allocation of all *N* nodes. The de-allocation forces in particular the termination at every node which was still attempting matches. The rationale is of course to avoid useless billing. Details depend on actual clouds. Once *E* has *s*, it recovers *S* as s_0 XOR s_1 and sends it to *X*.

Discussion

The scheme is correct and safe, [14]. In short, it is safe, since, at present, for any *N*, there is no way to disclose s_0 faster than through the RE_{NS} match attempts. Next, no cloud intruder can disclose S since E kept s_0. The scheme is correct, since it always recovers s_0. Especially, since the probability of a collision of a good one way hash is zero in practice. Also, it does not miss R provided that every T_i is that determined by C, i.e., the cloud is homogenous. Finally, as we show more below, the average recovery timing is $R/2$.

The scheme realizes the perfect static hash partitioning over a homogenous cloud. It generates the smallest cloud for any given R, provided that the in-cloud messaging time is negligible with respect to R. This leads to $\Delta \cong 0$ and should be the usual case. The average recovery time, say R^a is then $R / 2$, since every noise share is equally likely to be the noised one. Also the *Termination* interrupts every, but one node after the time $R/2 + \Delta$. See [14] for more on all the discussed issues.

In practice, even in a homogenous cloud, T_i's may little differ. The effect is some α variation, as for a not completely uniform hash. Also, for a short R and large N, messaging time may be not negligible. One direction towards respect of R, is then to calculate N, using $\alpha < 1$ in consequence, i.e., to have $\Delta > 0$. The theory of (static) hash files hints to $\alpha = 0.8 \div 0.9$ as promising choice. The probability of an overflow is then negligible under good hash assumption. N would in turn increase by 10-20%. We leave deeper study for the future.

Example 3. Continuing with Example 2, we consider the use of CloudLayer, [7]. The hourly pricing in Jan. 2012 guarantying the exclusivity of a node to the renter was 0.30c. This exclusivity called *private* option is crucial for cloud homogeneity. On this basis, we little extrapolate the pricing scheme, considering that 5m is the minimal charge per node used (1h at present). Since R was 10m, C may provide the cloud bill estimate as up to 400$ and 200$ on the average. This is generally already easily affordable. It could be further subject of insurance service. Costing then the insured key owner really peanuts, e.g., maybe 5$/y, if 1% users loose a key yearly.

Scalable Partitioning

We now suppose that every node n may *split* as we detail below. It is also the only to know its B, noted B_n. The scheme goes through the same phases as the static one. The *Init* phases are identical. In *Map* phase, node 0 also determines its T, now noted T_0, its *initial* load $L_0^{\mathrm{I}} = M / T_0$, its capacity $B_0 = R / T_0$ and its *initial* load factor $\alpha_0^{\mathrm{I}} = L_0^{\mathrm{I}} / B_0$. If by any chances, $\alpha_0^{\mathrm{I}} \leq 1$, node 0 executes the calculus, sends the result to E and terminates. Otherwise, as usual for files, we say that node 0 *overflows*. It then splits. Node 0 initiates then the *node level* j, as $j = 0$. Next, node 0 requests a new node that it labels node 2^j, i.e., node 1 here. Finally, it sets j to $j = 1$ and sends to node 2^j, the data (S'', j, a_0). Here, a_0 is again some (physical) address of node 0, letting the new node to send data back. From now on, the cloud has $N = 2$ nodes. Observe that the initial cloud had in fact $N = 2^0$ nodes for $j = 0$ at node 0 (alone). Now it scaled up to $N = 2^1$ nodes, with $j = 1$ at both nodes.

Next node 0 recalculates L_0 as $M / (2^j\ T_0)$ and α_0 as L_0 / B_0 accordingly. The reason is that *Reduce* phase if started for $N = 2$, would lead to $M / 2$ attempts for node 0 at max. Node 1 would obviously have to process $M / 2$ attempts as well. Iff $\alpha_0 > 1$, i.e., node 0 overflows still, it splits again. Otherwise, it starts *Reduce* phase. To split, node 0 creates new node and labels it node 2. This is, in fact, the result of the new (logical) node address computation as $0 + 2^j = 2$. Finally, node 0 performs $j := j + 1$ and sends (S", j, a_0) to node 2. Node 2 stores j as its own initial level. Hence, both nodes end up at level $j = 2$.

Node 1 acts similarly, starting with the measure of T_1, then of α_1. The node it creates, iff $\alpha_1 > 1$, is now node 3, as $3 = 1 + 2^j$ for $j = 1$ that was set upon node 1 creation. Both nodes end up at level $j = 2$ after $j := j + 1$ at node 1, like before. If both, node 0 and 1 split, the cloud reaches 2^2 nodes.

From now on, every node n that did not entered *Reduce* phase, recursively loops over the overflow testing and splitting. The general rule they act upon, compatible with the above, is as follows.

1. Every node n ; $n = 0,1$... after its creation or last split, calculates its L as $L_n = M / (2^j T_n)$ and tests whether $\alpha_n > 1$. For this purpose, a newly created node n, starts with the prediction of T_n. If there is no overflow, the node goes to *Reduce* phase.
2. Node splits. It creates node $n' = n + 2^j$, sets j to j+1 and sends (S'', j, a_0) to node n'.

In this way, for instance, as long as node 0 overflows, it continues to split, creating successively nodes 2^j with j starting at $j = 0$, i.e., nodes $n = 1, 2, 4, 8$... Node 1 starts with $j = 1$ and appends nodes $1+2^j$ that are $n = 3, 5, 9, 17$... Likewise node 2 starts with $j = 2$ and creates nodes $n = 2 + 2^j = 6, 10$... Every split appends a new node to the cloud. Each node has a unique logical address. Iff all the splitting nodes end up with the same j, these addresses form a contiguous space 0, 1...N-1 where $N = 2^j$. This happens for the homogenous cloud, i.e. same T everywhere. Finally, observe that for every splitting node n, every split halves L_n, hence α_n. The new node n' starts with L_n. But not necessarily with $\alpha_{n'} = \alpha_n$ as $T_{n'}$ and T_n may differ.

Every non-overflowing node n, i.e. with $\alpha \leq 1$, starts *Reduce* phase. There, it attempts matches of $H(f + k)$ for every $k \in$ [0, M-1] and such that $n = k \bmod 2^j$. Thus, for $j = 2$ for instance, node 0 loops over k =0, 4, 8... Likewise, node 2 for j = 1 loops over 2,6,10.... If node 1 did not split even once, hence it carries $j = 1$ and loops over the remaining values of [0, M[which are k =1,3,5,7... If the match succeeds at any node n > 0, i.e., $h = H(s)$ for some k and $s = f + k$, then node n sends s as s_0 to node 0 and exits. Otherwise, the match must succeed at node 0. Node 0 sends s_0 finally to E and starts the *Termination* phase. If no match succeeds at node n, node n also enters this phase locally. End result is every node gets de-allotted as for the static scheme. Details are up to the actual cloud in use.

Example 4. We continue with the previous data. We have thus $\alpha_0 = 2^{42\text{-}29} = 2^{13}$, initially. Node 0 splits 13 times till $\alpha_0 = 1$. The splits append nodes 1, 2, 4...4096. Then, node 0 enters *Reduce* phase. It attempts the matches for noises 0, 8192...2^{42} - 2^{13}. The nodes it has created split if needed and enter each Reduce phase as well. Likewise, node 1 has $\alpha = 2^{12}$ initially. It splits consequently 12 times, appending nodes 3, 5...4097. It then also enters *Reduce* phase. It attempts the matches for the noises 1, 8193...$2^{42} - 2^{13} + 1$. Its children split perhaps and also enter *Reduce* phase. Each scans "its" part of the noise space. And so on, for node 2, 3...4095. Node 4095 has $\alpha = 2$ initially. Hence it splits only once and we get $N = 2^{13}$. This is the optimal (size) cloud here, with every $\alpha = 1$. N is as small as for the static scheme. Notice that such α happens for the scalable one, any time we have $\alpha_0 = 2^i$ and all T's equal.

Consider now the same initial data for node 0. It would split again as above. Suppose in contrast for node 1 that $T_1 = 8T_0$, e.g., it is an 8-core node, compared to 1-core. Its capacity B increases eight fold as well. Then, we have initially $\alpha_1 = 2^9$ only. Node 1 would split only nine times, generating nodes 3, 5…513. Nodes 1025, 2049 and 4097 would not exist. The cloud would have at least three nodes less. It would be smaller than for the static scheme. *Reduce* phase at node 1 would process noises 1, 1025, 2049, 3073, 4097…

Finally, consider node 0 eight times faster than node 1. The splits of the latter would generate three more nodes. The static scheme would just fail to guarantee R.

Discussion

As for the static scheme, the scalable one is correct and safe. Tedious details are in [14]. While the scalable scheme aims on clouds where the static one would fail, it applies to homogenous clouds as well. Unlike in the special case above, however, since any $\alpha_n \in]0.5, 1]$ is equally likely then, average α should be about $\alpha = 75$ %. The average cloud size should increase by about $1/3N$, i.e., by about 30%. The worst case α is almost 50 %. This increases N almost twice with respect to the static scheme. Lower load factor leads in turn to faster match, under R and $R/2$ respectively. More precise determination of α and N in various conditions is a future goal.

Choosing a scalable scheme on a public cloud may remain advantageous despite possibly a larger cloud. The reason is the pricing structure. On CloudLayer for instance, [7], it should be in practice necessary to choose the already discussed private option for a static partitioning. The usual allocation mode termed *public* may lead to heterogeneity. However, the public mode is three times cheaper than the private one. Choosing a public option with scalable partitioning may then anyhow save somewhere between 1/3 and 2/3 of the bill. In sum, the scalable scheme is generally more versatile and is likely to be often preferred.

3 Related Work

The RE idea appeared apparently in [12]. It was applied to the client-side encrypted data outsourcing to a cloud. The outsourced data formed an LH^*_{RE} file distributed over cloud nodes. As its name suggests it is a Linear Hash based scalable distributed data structure (SDDS) [15], 2]. The encryption key was backed up as the secret with shares randomly distributed among the LH^*_{RE} encrypted data records. To recover the key, one needed to collect all the shares. A duly authorized cloud client could do it using the LH^*_{RE} scan operation. The cloud intruder in contrast had to break into typically many cloud nodes. Such break-ins appeared unlikely.

With respect to RE_{NS} schemes, the difference was the necessity of a multi-node cloud for the backup storage. Also, the scheme was designed only for the encryption keys of LH* files. These constraints do not exist for RE_{NS} schemes. One can see then a safe outsourcing of any SDDS file together with its RE_{NS} key.

The scheme termed CSCP in [13] also provides the client-side encrypted cloud outsourcing. Unlike for LH^*_{RE}, CSCP keys are however shared among authorized clients. CSCP uses a variant of the Diffie-Helman (DH) protocol for client authentication.

A private DH number constitutes the ID and lets the client to share or backup any key. Its loss inhibits these capabilities and may lead to key loss. To offset the danger, each private ID is backed up with admin. This may frustrate users of course. The RE_{NS} backup appears to nicely blend with CSCP scheme. Same observation applies to the popular SharePoint, using DH similarly for group messaging.

The RE concept roots also somehow in the classical cryptography concepts of *one-way hash with trapdoor* and of *cryptograms* or *crypto puzzles*. The Web provides numerous related references, e.g., [6]. RE may be seen as a one way hash where the cloud constitutes a distributed scalable trapdoor. Also, one may see RE as a scheme for cryptograms (crypto puzzles) that the owner makes complex at will through the quantity of noise injected. At the same time, the decryption algorithm makes the resolution scalable and distributed also somehow at will. This leads to perhaps other RE schemes exploring ideas in these domains.

The risks of key escrow (a.k.a. key archive, key backup, key recovery) we have mentioned are hardly a new issue. They were studied for decades, especially intensively in the late nineties, [1, 8, 17]. The wave was triggered by the famous Clipper chip envisioned for storing cryptographic keys by government or private agencies. The concept of recoverable encryption in this context was somehow implicit in the taxonomy in [8], becoming explicit in a revision of that work [9]. However, our recoverability is basically an optional service. The initial meaning was mandatory, for the law enforcement, [10].

That idea was criticized for inherent dangers and overreach. As an alternate, Shamir proposed government key escrow using only partial keys. A government agency could recover the key by making a brute force attack on the unknown part of the key feasible. Rivest & al proposed in turn non-parallelizable crypto-puzzles, for access after a guaranteed minimum amount of time only, [16]. It was observed also that key escrow for the benefit of someone else may need to verify that the escrowed key is indeed the true key [5]. This does not apply to our case, since there is no adversarial relation between the beneficiary of key recovery and the data owner. Finally, Blaze proposed a yet different approach, through a massive secret sharing scheme distributing shares of escrowed keys to many (key) share-holders [3]. None of these techniques was scalable or was even meant to be.

With respect to illegal attempts, botnets are a temptation avoiding legitimate clouds all together. This danger seems however increasingly remote. Many anti-botnet tools appeared. Most computers have antivirus protection. Internet users exercise caution. Finally, botnets became widely recognized as a cybercrime.

4 Conclusion

Encryption key safety is the Achilles' heel of modern cryptography. Key backups are necessary. Copies increase however in turn the disclosure risk. Our RE_{NS} schemes alleviate the dilemma. The key remains always recoverable provided the cloud large enough for the key owner to deter unwelcome curiosity. We believe our proposals to have many applications.

Future work should analyze the schemes more in depth. Experiments are also a necessity. There are also variants, above and those presented in [14]. The one called *multiple noising* appears of particular interest. Finally, the key insurance service idea seems with high potential.

Acknowledgments. Our schemes result in part from helpful discussions with Jonathan Katz (U. of Maryland, College Park). Sushil Jajodia was partially supported by NSF under grants CT-20013A and CCF-1037987.

References

1. Abelson, H., Anderson, R., Bellovin, S.M., Benaloh, J., Blaze, M., Gilmore, J., Neumann, P.G., Rivest, R.L., Schneier, B.: The Risks of Key Recovery, Key Escrow, and Trusted Third-Party Encryption, `http://www.crypto.com/papers/escrowrisks98.pdf`
2. Abiteboul, S., Manolescu, I., Rigaux, P., Rousset, M.C., Senellart, P.: Web Data Management. Cambridge University Press (2011)
3. Blaze, M.: Oblivious Key Escrow. In: Anderson, R. (ed.) IH 1996. LNCS, vol. 1174, pp. 335–343. Springer, Heidelberg (1996)
4. Crypto++ 5.6.0 Benchmarks, `http://www.cryptopp.com/benchmarks.html`
5. Bellare, M., Goldwasser, S.: Verifiable partial key escrow. In: 4th ACM CCS Conf., pp. 78–91 (1997)
6. Chandrasekhar, S.: Construction of Efficient Authentication Schemes Using Trapdoor Hash Functions. Ph.D Dissertation. University of Kentucky (2011)
7. CloudLayer Hourly Pricing, `http://www.softlayer.com/cloudlayer/computing/`
8. Denning, D.E., Branstad, D.K.: A Taxonomy for key escrow encryption systems. Communications of the ACM 39(3) (1966)
9. Denning, D.E., Branstad, D.K.: A Taxonomy for Key Recovery Encryption Systems, `http://faculty.nps.edu/dedennin/publications/TaxonomyKeyRecovery.htm`
10. Lee, R.D.: Testimony of Ronald D. Lee, Attorney General.... (March 1999), `http://www.cybercrime.gov/leesti.htm`
11. Miller, E., Long, D., Freeman, W., Reed, B.: Strong security for distributed file systems. In: Proceedings of the Conference on File and Storage Technologies (FAST 2002), pp. 1–13 (January 2002)
12. Jajodia, S., Litwin, W., Schwarz, T.: LH*$_{RE}$: A Scalable Distributed Data Structure with Recoverable Encryption. In: IEEE-CLOUD 2010 (2010)
13. Jajodia, S., Litwin, W., Schwarz, T.: Privacy of Data Outsourced to a Cloud for Selected Readers through Client-Side Encryption. In: CCS 2011 Workshop on Privacy in the Electronic Society (WPES 2011), Chicago (2011)
14. Jajodia, S., Litwin, W., Schwarz, T.: Recoverable Encryption Through Noised Secret. Electronic Res. Rep. (2011), `http://www.lamsade.dauphine.fr/~litwin/Recoverable%20Encryption_10.pdf`
15. Litwin, W., Neimat, M.-A., Schneider, D.: LH* - A Scalable Distributed Data Structure. ACM TODS 12 (1996)
16. Rivest, R.L., Shamir, A., Wagner, D.A.: Time-lock puzzles and timed-release crypto. Technical Report, Massachusetts Institute of Technology, MIT/LCS/TR-684 (1996)
17. Walker, S., Lipner, S., Ellison, C., Balenson, D.: Commercial key recovery. Communications of the ACM 39 (March 3, 1996)

A Bigtable/MapReduce-Based Cloud Infrastructure for Effectively and Efficiently Managing Large-Scale Sensor Networks

Byunggu Yu[1], Alfredo Cuzzocrea[2], Dong Jeong[1], and Sergey Maydebura[1]

[1] Department of Computer Science and Information Technology
University of the District of Columbia, Washington, DC, USA
{byu,djeong}@udc.edu, dc.2007@yahoo.com
[2] ICAR-CNR and University of Calabria
Rende, Cosenza, Italy
cuzzocrea@si.deis.unical.it

Abstract. This paper proposes a novel approach for effectively and efficiently managing large-scale sensor networks defining *a Cloud infrastructure that makes use of Bigtable at the data layer and MapReduce at the processing layer*. We provide principles and architecture of our proposed infrastructure along with its experimental evaluation on a real-life computational platform. Experiments clearly confirm the effectiveness and the efficiency of the proposed research.

Keywords: Bigtable, MapReduce, Cloud computing, Large-scale sensor networks.

1 Introduction

An increasing number of emerging applications deal with a large number of *Continuously Changing Data Objects* (CCDOs). CCDOs [10], such as vehicles, humans, animals, mobile sensors, nano-robots, orbital objects, economic indicators, geographic water regions, forest fires, risk regions associated with transportation routes, sensor data streams, and bank portfolios (or assets), range from continuously moving physical objects in a two-, three-, or four-dimensional space-time to conceptual entities that can continuously change in a high-dimensional information space-time. A variety of handheld computers and smartphones equipped with a GPS and other sensory devices are already available to consumers. A combined sensor that can detect and report $n \geq 1$ distinct stimuli plots a sequence of data points in the (n+1)-dimensional data space-time (i.e., n data dimensions and one time dimension). In GPS applications, the locations of objects can continuously change in the geographic space. In earth science applications, temperature, wind speed and direction, radio or microwave image, and various other measures associated with a certain geographic region can change continuously. In particular, for what regards to images, they can be mapped onto multidimensional points (i.e., vectors) through

A. Hameurlain et al. (Eds.): Globe 2012, LNCS 7450, pp. 25–36, 2012.

principal components analysis. In this sense, a sequence of images or video clip associated with a certain ontological entity is a CCDO. In water-resource research, the flow, pH, dissolved oxygen, salinity, and various other attributes can change continuously at each monitoring site. More and more CCDO applications are appearing as the relevant technologies proliferate. There is much common ground among various types of CCDOs: CCDOs produce discrete points representing the breadcrumbs of their trajectories in the information space-time.

Recent advances and innovations in smart sensor technologies, energy storage, data communications, and distributed computing paradigms are enabling technological breakthroughs in very large sensor networks. There is an emerging surge of next-generation sensor-rich computers in consumer mobile devices as well as tailor-made field platforms wirelessly connected to the Internet. Very large sets of CCDOs are growing over time, posing both challenges and opportunities in relation to scalable and reliable management of the peta- and exa-scale time series being generated and accumulated over time.

This clear trend poses several issues from the data management point-of-view, as traditional models, algorithms and schemes (mostly developed in the context of *distributed database systems*) cannot be applied as-they-are to the challenging context of managing massive CCDOs generated by large-in-size sensor networks. On the other hand, recent *Cloud technologies* offer several optimization opportunities that may be used to this end. Cloud computing is a successful computational paradigm for managing and processing big data repositories, mainly because of its innovative metaphors known under the terms "*Database as a Service*" (DaaS) [11] and "*Infrastructure as a Service*" (IaaS). DaaS defines a set of tools that provide final users with seamless mechanisms for creating, storing, accessing and managing their proper databases on remote (data) servers. Due to the naïve features of big data like those generated by CCDOs, DaaS is the most appropriate computational data framework to implement big data repositories [12]. *MapReduce* [5] is a relevant realization of the DaaS initiative. IaaS is a provision model according to which organizations outsource infrastructures (i.e., hardware, software, network) used to support ICT operations. The IaaS provider is responsible for housing, running and maintaining these services, by ensuring important capabilities like *elasticity*, *pay-per-use*, *transfer of risk* and *low time to market*. Due to specific application requirements of applications running over big data repositories, IaaS is the most appropriate computational service framework to implement big data applications. *Bigtable* [4] is the *compressed database system* running over MapReduce and other Google systems like *Google Maps*. It is deployed over a distributed file system setting.

This paper presents a Cloud-computing approach to this issue based on the following well-known data storage and processing paradigms: Bigtable [4] and MapReduce [5], along with its experimental assessment. The remaining part of the paper is organized as follows. Section 2 provides an overview on sensor-network data management proposals appearing in literature. The conceptual basis of CCDO is discussed in Section 3. Section 4 presents a Bigtable model designed for very large sets of CCDOs generated by sensor networks. Section 5 focuses the attention on the physical implementation of our Bigtable model. Section 6 discusses scalability issues of the proposed Bigtable approach. In Section 7, we provide the experimental evaluation and analysis of our proposed approach. In Section 8, we provide and discuss open problems and actual research trends related to the issue of managing very large sensor-network data over Bigtable. Finally, Section 9 concludes the paper and proposes future work to be considered in this scientific field.

2 Related Work

The problem of efficiently managing massive sensor-network data generated by large-in-size sensor networks has received significant attention during the last decade [20]. In [21], *compression paradigms over Grids* are exploited to obtain efficiency during indexing and querying actives. In line with this, [22] further extends this work towards a more effective *event-based lossy compression*, with OLAP [23] querying capabilities. OLAP is also a way to compress sensor-network data via *multidimensional summarization* [24]. *Uncertainty* and *imprecision* are also significant aspects of sensor-network data processing. A solution is discussed in [25].

Traditional and well-recognized proposals in the wide sensor-network data management context are the following. *Cougar* [16] is an approach that views the sensor-network data management problem in terms of a *distributed database management problem*, and supports distributed query processing over such a database. *TinyDB* [17] is an *acquisitional and aggregate query processing system* over large-scale sensor networks that also ensures energy efficiency. To this end, TinyDB relies on sophisticated *probabilistic models*. *BBQ* [18] is a further refinement of TinyDB where probabilistic models are enhanced via elegant *statistical modeling techniques*, even with *approximation metaphors*. *PicoDBMS* [19] is a database platform powered by features that allow *smart-card data* (which clearly resemble sensor-network data) to be managed efficiently.

Among more recent initiatives, we recall: [13], which studies the challenges of managing data for the *Sensor Web*, and defines open issues in this field; [14], which proposes the system *StonesDB* that support indexing and management capabilities for sensor-network data via well-known *flash memory* technologies; [15], which presents *PRESTO*, a two-tier sensor data management architecture that comprises proxies and sensors *cooperatively acting* to support acquisition and query processing functionalities.

With respect to the mentioned related work, our approach is innovative as we propose using Bigtable to manage large-scale sensor network data directly, hence achieving effectiveness and efficiency in this critical activity.

3 A Generic Ontology for Modeling CCDO Objects

Sooner than later, more complex and larger applications that deal with higher-dimensional CCDOs (e.g., a moving sensor platform capturing multiple stimuli) will become commonplace – increasingly complex sensor devices will continue to proliferate alongside their potential applications.

To support large-scale CCDO applications, one requires a data management system that can store, update, and retrieve CCDOs. Importantly, although CCDOs can continuously move or change (thus drawing continuous trajectories in the information space-time), computer systems cannot deal with continuously occurring infinitesimal changes. Thus, each object's continuously changing attribute values (e.g., position in the information space and the higher-order derivatives of the changing positions) can only be discretely updated. Hence, they are always associated with a degree of uncertainty.

Considering an observer who records (or reports) the state of a continuously moving object as often as possible, the object is viewed as a sequence of connected segments in space-time, and each segment connects two consecutively reported states of the object. Hence, an *Ontology*[3]-based CCDO concept can be modeled in terms of an abstracted multi-level point-type CCDO concept, which is composed by the following elements (a more detailed description can be found in [10]):

- *Trajectory.* A trajectory consists of dynamics and can be modeled as a function *f*: *time* → *snapshot*, where *time* is a past, current, or future point over time.
- *Snapshot.* A snapshot is a probability distribution that represents the (estimated) probability of every possible state in the data space at a specific point over time. Depending on the dynamics and update policies, the probability distribution may or may not be bounded.
- *State.* A state is a tuple: $\langle P, O \rangle$, where P is a location (position) in the data space-time (i.e., a point in the space-time), and O is an optional property list including zero or more of the following: orientation, location velocity vector (and rotation velocity vector, if orientation is used), location acceleration vector (and rotation acceleration vector, if orientation is used), and even higher derivatives at the time of P.
- *Dynamics.* The dynamics of a trajectory is the lower and upper bounds of the optional properties of all states of the trajectory.

For any CCDO that draws a continuous trajectory in space-time, only a subset of the object's states can be stored in the database. Each pair of consecutively reported states of the CCDO represents a single trajectory segment. Any number of unknown states can exist between the two known states. In order to support queries referring to the trajectories without false dismissals, one needs an estimation model, called the *uncertainty model* that covers all possible unknown states (conservative estimation of all possible unknown states).

Spatio-temporal uncertainty models reported in [6,8,9] formally define both past and future spatio-temporal uncertainties of CCDO trajectories of any dimensionality. In these next-generation models, the uncertainty of the object during a time interval is defined to be the overlap of two spatio-temporal volumes, called funnels or tornados. A higher-degree model presented in [9] substantially reduces the uncertainty by taking into account the temporally varying higher-order derivatives, such as velocity and acceleration.

4 The CCDO Bigtable Proposal

This Section proposes our Bigtable-based approach to the problem of managing very large sets of time-growing CCDOs. A *CCDO Bigtable* is a Bigtable schema designed for very large sensor networks collectively generating peta- and exa-byte data over time. This paper proposes a modified version of the Bigtable model constructs by means of the following components:

- *Row Key*. It is a unique ID of the individual CCDO.
- *Column Keys*. They can be of the following column families [4]:
    - *CCDO.description*. It is an XML description of the CCDO. This column family may have only one column key without qualifiers. This family represents the most sparse column family of the CCDO Bigtable.
    - *Trajectory.dynamics*. This family may have multiple column keys with low- to high-order dynamics as the qualifiers. The qualifier format is of kind: *sensor_name.time_series_name. dynamics_name*, where *sensor_name* is the logical identifier of the target sensor, *time_series_name* is the logical identifier of the reference time series, and *dyamics_name* is the logical identifier of the trajectory dynamics. Examples of value for this family are: Trajectory.dynamics:gps.longitude.max_velocity, Trajectory.dynamics:accelerometer.y.max_acceleration, Trajectory.dynamics:temperature.max_rate_of_change.
- *Reported_State*. This family has a distinct column key for each distinct sensor time series. It may represent the least sparse among all the families. The qualifier format is of kind: *sensor_name.time_series_name*, where *sensor_name* is the logical identifier of the target sensor, and *time_series_name* is the logical identifier of the reference time series. Examples of value for this family are: Reported_State:gps.longitude, Reported_State:accelerometer.y, Reported_State:temperature.

Note that, in a CCDO Bigtable, each entry (or cell) can contain multiple versions of the same data or time series with timestamps.

Actually, *Apache Hadoop* [1] and *Apache HBase* [2] are the most prominent and popular open source implementations of the Google's proprietary Cluster technology and Bigtable, respectively. Through the rest of this paper, we assume the Hadoop and HBase environment as the reference computational environment.

Let us now focus on more details on Bigtable. A Bigtable is a sorted table. The key structure comprises row key (most significant), column family, qualifier, and timestamp (least significant and sorted in reverse). Each ⟨*row_key*, *column_key*, *timestamp*⟩ exactly specifies a certain cell value at a certain point in time, and having an unbounded number of time-stamped values of the same cell is possible. As of this writing, HBase recommends less than or equal to three column families and abbreviated row key and column key names [2].

A growing time series can be stored in a single cell via the built-in "versioning" (by using the attribute *HColumnDescriptor*). Alternatively, one can adopt "rows approach" [2], in which the timestamp becomes part of the row key in each column family. For example, one can add a timestamp suffix to the row key of the CCDO Bigtable, in our implementation. In this alternative approach, the maximum number of different updates that can happen at the same time in the same column family should be defined for each column family set as the maximum number of versions in the attribute *HColumnDescriptor*. Yet another alternative is to add the timestamp as suffix to the column qualifier. This does not incur any versioning issues. However, it is important to note that different approaches will result in different significance of the timestamp in the key order and carefully programmed temporal logic in the application program will be required.

The operation get implemented in HBase allows a direct access, given a table name, a row key and optionally a dictionary (i.e., description) of values or ranges of subsequent column keys and timestamps. The operation scan allows a table access by an ad hoc and arbitrary combination of selected column family names, the qualifier names, timestamp, and cell values of the table by using the class *Filter*.

Our previous work reported in [10] presents a spatio-temporal paradigm for dealing with the uncertainties of sensor data streams from a database management perspective with newly found insights into the computational efficiency: a MapReduce approach for parallelizing the uncertainty computation for enhanced performance, resolution, and scalability, in the relational setting. The MapReduce-based solution in [10] can be applied to the CCDO Bigtable presented in this paper in a straightforward manner as well.

5 Physical Architecture Overview

HBase runs on a networked cluster of computers powered by the *Hadoop Distributed File System* (HDFS) [1] and parallel job scheduling (via MapReduce). HDFS is a proprietary Google file system solution for supporting file-oriented, distributed data management operations efficiently. HBase system consists of a master and one or more region servers running on the cluster. Every HBase table (Bigtable) consists of one or more regions distributed among the region servers.

Each region consists of one or more stores each of which is a physical representation of a distinct column family. In our case, the CCDO Bigtable, each region consists of three stores for the three column families (see Section 4): *CCDO.description*, *Trajectory.dynamics*, *Reported_State*. The row key attribute (see Section 4) is associated with each column family in the corresponding store. This way, empty cells are not stored in the physical stores.

Each store includes an allocated memory space, called *memstore*, of the region server and one or more data files called *StoreFiles*. Each incoming write to the table, having *Put*, *Delete*, *DeleteColumn*, or *DeleteFamily* as key type, is first written in the corresponding region's *memstore*. When *memstore* becomes full, all contents are written as a new immutable *StoreFile* in the distributed file system. Hadoop compaction process merges the immutable *StoreFiles* into a new immutable *StoreFile*, during which multiple writes with different key types regarding the same cell are resolved (deletes are performed only by major compaction running on a regular basis and combining all *StoreFiles* of the store into a single *StoreFile*), and deletes the old *StoreFiles*.

A region split occurs whenever there is a *StoreFile* larger than a predefined threshold (default 256MB). When a compaction produces such a large *StoreFile*, the region (i.e., every store of the region) is split into two. Since each region spans all column families as stores, each split is across the entire width of the table.

As regards indexing aspects, which play a critical role in this context, all records of a Bigtable are in a sorted order by the multiple key: {*row key*, *column family qualifier*, *time stamp*}, with the timestamp being in reverse order (i.e., the most recent first).

Hence, the physical *StoreFile* is an indexed file (the *HFile* format in Hadoop [1]) with the index portion loaded in the main memory. This solution efficiently supports column value accesses [4].

6 Remarks on Scalability Issues

Once the number of regions has started increasing by region splits, the system starts distributing the regions among different region servers for load balancing. Moving a region from one computer of the cluster to another requires locality consideration. The region server will show the best performance when the region is available on its local disk partition representing a portion of the distributed file system space.

This physical move of a region from one node to another happens via the locality-prioritized block writing capability of HDFS. A region server may be assigned a region with non-local *StoreFiles*. When a new *StoreFile* (from the *memstore* or a compaction) is being created, the file blocks are written in the distributed file system. The distributed file system is a collection of the assigned partitions of the disks of the cluster node computers. For each block produced by any region server, the system writes the first replica on the local disk of the region server and the second replica on a disk nearby in the network topology (in the same rack) and the third replica in a different rack (for a better node fault tolerance). Over time, through the compaction process, the region data will become local to the region server, eventually achieving the region-region server locality.

It has been found that this design provides rather balanced (uniform) distribution of regions (data) across the region servers. This can provide almost linear improvement regarding the table data read/write performance with increasing number of cluster nodes and almost constant with increasing data size [4].

Given a table access (e.g., get or scan operations with filters), a map process (mapper) is created for each region. Multiple mappers can access their regions of the CCDO Bigtable at the same time, resulting in a parallel processing of the query over the distributed regions. The actual degree of the parallelism is naturally limited by the size of the cluster (in terms of the number of computing nodes).

7 Experimental Evaluation and Analysis

To better understand and study the performance of the proposed Bigtable approach, we conducted an experimental campaign on a Hadoop Cloud. The main focus of our experimental campaign was on assessing how the data insertion performance changes as the CCDO data rapidly grows over time, which is a typical applicative setting of real-life large-scale sensor networks. In these experiments, *MySQL* was used as a baseline benchmark-platform.

To implement and test the proposed Bigtable approach, we engineered a Hadoop Cloud consisting by 1 master node and 25 slave nodes. The master was a dual-core desktop with 2GB memory, and each slave was a single-core desktop with 1GB memory. Hadoop stable version 1.0.2 and HBase stable version 0.92.0 were adopted, respectively. All configurations were set to the default. The *Apache ZooKeeper* [27] service was setup with 1 server, i.e. the master node. The *durable sync* mode of

Hadoop-HBase was on. The master was also used as the HBase master node and a single region server instance was created on each individual slave node. Separately, we set a MySQL machine with four times more cores and memory than any of the slave nodes.

The test data set was created based on 1,000,000 unique *Row Key* values. For each CCDO (unique *Row Key*), one *CCDO_description* and three sets of *Trajectory.dynamics* of 6 values each (6 distinct qualifiers) were created. For each set of *Trajectory.dynamics*, 1,000 *Reported_state* sets were created. Each *Reported_state* set consisted of 10 distinct state values (10 distinct qualifiers). Strictly based on the relational CCDO database design presented in [10], which comprises *CCDO* table, *Trajectory* table, and *Reported_state* table, a relational database was created on the MySQL machine, as comparison benchmark. The test data resulted in 3,004 INSERT statements for each unique CCDO: 1 row insert into the *CCDO* table; 3 row inserts into the *Trajectory* table; and 3,000 row inserts into the *Reported_state* table.

```
Module Sequential CCDO Insertion

Get ID from the raw data file;
Insert ID description
Repeat{ //3 times
        Insert dynamics; //6 values
        Repeat{ //1,000 times
                Insert reported state; //10 values
                Increase timestamp;
        }
}
```

Fig. 1. The sequential execution version for CCDO insertion module

```
Module Parallel CCDO Insertion

Get a line from the raw data file; //64MB
Parse the line into string-array using tab-separate components;
Map phase{
        Two-component line task: Insert the new Row Key value:
        Three-component line task: Insert the corresponding column value;
}
Reduce phase
```

Fig. 2. The sequential execution version for CCDO insertion module

The Bigtable counterpart was created on the Hadoop-HBase Cluster as an HBase table. In the Bigtable approach, each unique CCDO translated into 30,019 PUT statements: 1 (description) + 3 × 6 (trajectory dynamics) + 3 × 1,000 × 10 (reported states). The versioning (with max) was used for generating timestamps. The Bigtable approach has been implemented according to different two versions of CCDO insertion modules: *sequential version* and *parallel version*, respectively. The parallel version was a MapReduce version, utilizing as many mappers as possible to improve the insertion rate. The sequential version created one mapper only, whose behavior is described by the pseudo-code listed in Fig. 1. Fig. 2 instead lists the pseudo-code describing the behavior of the parallel version with multiple-mappers.

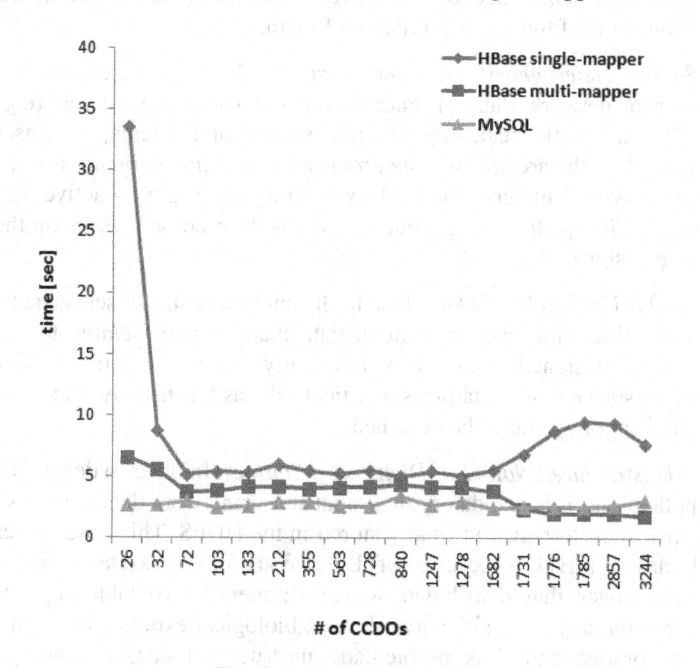

Fig. 3. Experimental results

Fig. 3 shows the effective growing rates of data population as the CCDO object database rapidly grows over time. Here, performance of the single-mapper approach (i.e., the sequential version of the Bigtable approach), the multiple-mapper approach (i.e., the parallel version of the Bigtable approach), and the MySQL-based approach in terms of amount of time (in seconds) necessary to one full CCDO insertion constituted by 30,019 PUT statements for the Bigtable approach (in contrast to 3,004 INSERT statements for the MySQL approach) are compared. From the analysis of Fig. 3, it clearly follows that the single-mapper approach improves quickly over the initial growth as the initial region is split and distributed over the region servers. However, this initial improvement stagnates after, as the single-mapper approach becomes a bottleneck itself. The multiple-mapper approach, instead, starts with distributed mappers and performance improves in a steady manner as the region servers are dynamically reorganized with high locality over time, and after becomes superior to the MySQL approach, which shows relatively flat rates. This analysis clearly confirms the effectiveness and the efficiency of the proposed research.

8 Managing Very Large Sensor-Network Data over Bigtable: Open Problems

There are a number of open problems and actual research trends related to the issue of managing very large sensor-network data over Bigtable. In the following, we provide an overview on some of the most significant of them.

(*a*) *Data Source Heterogeneity and Incongruence*. Very often, distributed sources producing sensor-network data of interest for the target application (e.g., legacy systems, Web, scientific data repositories, sensor and stream databases, social networks, and so forth) are *strongly heterogeneous and incongruent*. This aspect not only conveys in typical integration problems, mainly coming from active literature on *data and schema integration issues*, but it also has deep consequences on the kind of analytics to be designed.

(*b*) *Filtering-Out Uncorrelated Data*. Due to the enormous size of sensor-network data repositories, dealing with large amount of data that are *uncorrelated* to the kind of application to be designed occurs very frequently. As a consequence, filtering-out uncorrelated sensor-network data plays a critical role, as this heavily affects the *quality* of final Bigtable applications to be designed.

(*c*) *Strongly Unstructured Nature of Distributed Sources*. In order to design meaningful Bigtable applications, it is mandatory that input sensor-network data are transformed in a suitable, structured format, and finally stored in the HDFS. This poses several issues that recall classical ETL processes of Data Warehousing systems, but with the additional challenges that distributed sources alimenting Bigtable repositories are *strongly* unstructured (e.g., social network data, biological experiment result data, and so forth) in contrast with less problematic unstructured data that are popular in traditional contexts (e.g., XML data, RDF data, and so forth). Here, transformations from unstructured to structured format should be performed on the basis of the analytics to be designed, according to a sort of *goal-oriented methodology*.

(*d*) *High Scalability*. High scalability is one of the primer features to be ensured for a very large sensor-network data management system over Bigtable. To this end, exploiting the Cloud computing computational framework seems to be the most promising way to this end [12], like we do in our CCDO Bigtable based implementation. The usage of the IaaS-inspired Cloud computing computational framework is meant with the aim of achieving some important characteristics of highly-scalable systems, among which we recall: (*i*) *"true" scalability*, i.e. the effective scalability that a powerful computational infrastructure like Clouds is capable of ensuring; (*ii*) *elasticity*, i.e. the property of rapidly adapting to massive updates and fast evolutions of big data repositories; (*iii*) *fault-tolerance*, i.e. the capability of being robust to faults that can affect the underlying distributed data/computational architecture; (*iv*) *self-manageability*, i.e. the property of *automatically* adapting the framework configuration (e.g., actual load balancing) to rapid changes of the surrounding data/computational environment; (*v*) *execution on commodity machines*, i.e. the capability of scale-out on thousands and thousands of commodity machines when data/computational peaks occur.

9 Conclusions and Future Work

Inspired by recent trends in the context of effectively and efficiently managing very large sensor-network data, in this paper we have focalized on the issue of managing datasets modeling massive CCDOs generated by large-in-size sensor networks. We introduced an innovative Cloud-computing-based solution that combines two well-known data storage and processing paradigms, namely Bigtable and MapReduce, to gain effectiveness, efficiency and reliability during this so-engaging task. This has conveyed in the so-called CCDO Bigtable concept, which extends traditional Bigtable. We provided an Ontology-based conceptualization of CCDO Bigtable objects, along with its physical implementation, and discussed scalability issues related to so-defined objects. We also provided the experimental evaluation and analysis of our proposed Cloud infrastructure, which confirmed the benefits deriving from applying this infrastructure in modern Cloud computing systems for managing large-scale sensor networks.

Future work is mainly oriented towards improving the performance of search and retrieval capabilities in CCDO Bigtable, by means of a suitable empirical study. Also, we are currently integrating the CCDO Bigtable within the web-based and database-supported three-tier sensor cloud system *Smart Sensor Command and Control System* [26].

Acknowledgement. This research was supported by the U.S. National Science Foundation (NSF) Grants 0911969 and 0940393. The authors are grateful to Mr. Soufiane Berouel from the Department of Computer Science and Information Technology, University of the District of Columbia, Washington, DC, USA, for his contribution to the experimental evaluation and analysis of the proposed Cloud infrastructure.

References

1. Apache Hadoop, `http://hadoop.apache.org`
2. Apache, HBase, `http://hbase.apache.org`
3. Uschold, M., Gruninger, M.: Ontologies: Principles, Methods and Applications. Knowledge Engineering Review 11(2), 93–155 (1996)
4. Chang, F., Dean, J., Ghemawat, S., Hsieh, W.C., Wallach, D.A., Burrows, M., Chandra, T., Fikes, A., Gruber, R.E.: Bigtable: A Distributed Storage System for Structured Data. ACM Transactions on Computer Systems 26(2), Art. 4 (2008)
5. Dean, J., Ghemawat, S.: MapReduce: Simplified Data Processing on Large Clusters. Communications of the ACM 51(1), 107–113 (2008)
6. Yu, B.: A Spatiotemporal Uncertainty Model of Degree 1.5 for Continuously Changing Data Objects. In: Proceedings of ACM SAC Int. Conf., pp. 1150–1155 (2006)
7. Yu, B., Bailey, T.: Processing Partially Specified Queries over High-Dimensional Databases. Data & Knowledge Engineering 62(1), 177–197 (2007)
8. Yu, B., Kim, S.H.: Interpolating and Using Most Likely Trajectories in Moving-Objects Databases. In: Bressan, S., Küng, J., Wagner, R. (eds.) DEXA 2006. LNCS, vol. 4080, pp. 718–727. Springer, Heidelberg (2006)

9. Yu, B., Kim, S.H., Alkobaisi, S., Bae, W.D., Bailey, T.: The Tornado Model: Uncertainty Model for Continuously Changing Data. In: Kotagiri, R., Radha Krishna, P., Mohania, M., Nantajeewarawat, E. (eds.) DASFAA 2007. LNCS, vol. 4443, pp. 624–636. Springer, Heidelberg (2007)
10. Yu, B., Sen, R., Jeong, D.H.: An Integrated Framework for Managing Sensor Data Uncertainty using Cloud Computing. Information Systems (2012), doi:10.1126/j.is.2011.12.003
11. Hacigumus, H., Iyer, B., Mehrotra, S.: Providing Database as a Service. In: Proceedings of IEEE ICDE Int. Conf., pp. 29–38 (2002)
12. Agrawal, D., Das, D., El Abbadi, A.: Big Data and Cloud Computing: Current State and Future Opportunities. In: Proceedings of EDBT Int. Conf., pp. 530–533 (2011)
13. Balazinska, M., Deshpande, A., Franklin, M.J., Gibbons, P.B., Gray, J., Hansen, M.H., Liebhold, M., Nath, S., Szalay, A.S., Tao, V.: Data Management in the Worldwide Sensor Web. IEEE Pervasive Computing 6(2), 30–40 (2007)
14. Diao, Y., Ganesan, D., Mathur, G., Shenoy, P.J.: Rethinking Data Management for Storage-centric Sensor Networks. In: Proceedings of CIDR Int. Conf., pp. 22–31 (2007)
15. Li, M., Ganesan, D., Shenoy, P.J.: PRESTO: Deedback-Driven Data Management in Sensor Networks. IEEE/ACM Transactions on Networking 17(4), 1256–1269 (2009)
16. Yao, Y., Gehrke, J.E.: The Cougar Approach to In-Network Query Processing in Sensor Networks. SIGMOD Record 31(3), 9–18 (2002)
17. Madden, S., Franklin, M., Hellerstein, J., Hong, W.: TinyDB: An Acqusitional Query Processing System for Sensor Networks. ACM Transactions on Database Systems 30(1), 122–173 (2005)
18. Deshpande, A., Guestrin, C., Madden, S., Hellerstein, J.M., Hong, W.: Model-Driven Data Acquisition in Sensor Networks. In: Proceedings of VLDB Int. Conf., pp. 588–599 (2004)
19. Ganesan, D., Greenstein, B., Perelyubskiy, D., Estrin, D., Heidemann, J., Govindan, R.: Multi-Resolution Storage in Sensor Networks. ACM Tranasctions on Storage 1(3), 277–315 (2005)
20. Cuzzocrea, A.: Intelligent Techniques for Warehousing and Mining Sensor Network Data. IGI Global (2009)
21. Cuzzocrea, A., Furfaro, F., Mazzeo, G.M., Saccà, D.: A Grid Framework for Approximate Aggregate Query Answering on Summarized Sensor Network Readings. In: Proceedings of GADA Int. Conf., pp. 144–153 (2004)
22. Cuzzocrea, A., Chakravarthy, S.: Event-Based Lossy Compression for Effective and Efficient OLAP over Data Streams. Data & Knowledge Enginering 69(7), 678–708 (2010)
23. Gray, J., Chaudhuri, S., Bosworth, A., Layman, A., Reichart, D., Venkatrao, M., Pellow, F., Pirahesh, H.: Data Cube: A Relational Aggregation Operator Generalizing Group-by, Cross-Tab, and Sub Totals. Data Mining and Knowledge Discovery 1(1), 29–53 (1997)
24. Cuzzocrea, A.: CAMS: OLAPing Multidimensional Data Streams Efficiently. In: Pedersen, T.B., Mohania, M.K., Tjoa, A.M. (eds.) DaWaK 2009. LNCS, vol. 5691, pp. 48–62. Springer, Heidelberg (2009)
25. Cuzzocrea, A.: Retrieving Accurate Estimates to OLAP Queries over Uncertain and Imprecise Multidimensional Data Streams. In: Bayard Cushing, J., French, J., Bowers, S. (eds.) SSDBM 2011. LNCS, vol. 6809, pp. 575–576. Springer, Heidelberg (2011)
26. UDC Projects, `http://informatics.udc.edu/projects.php`
27. Apache ZooKeeper, `http://zookeeper.apache.org/`

MapReduce Applications in the Cloud: A Cost Evaluation of Computation and Storage

Diana Moise and Alexandra Carpen-Amarie

INRIA Rennes - Bretagne Atlantique / IRISA

Abstract. MapReduce is a powerful paradigm that enables rapid implementation of a wide range of distributed data-intensive applications. The Hadoop project, its main open source implementation, has recently been widely adopted by the Cloud computing community. This paper aims to evaluate the cost of moving MapReduce applications to the Cloud, in order to find a proper trade-off between cost and performance for this class of applications. We provide a cost evaluation of running MapReduce applications in the Cloud, by looking into two aspects: the overhead implied by the execution of MapReduce jobs in the Cloud, compared to an execution on a Grid, and the actual costs of renting the corresponding Cloud resources. For our evaluation, we compared the runtime of 3 MapReduce applications executed with the Hadoop framework, in two environments: 1)on clusters belonging to the Grid'5000 experimental grid testbed and 2)in a Nimbus Cloud deployed on top of Grid'5000 nodes.

1 Introduction

MapReduce has recently emerged as a powerful paradigm that enables rapid implementation of a wide range of distributed data-intensive applications. When designing an application using the MapReduce paradigm, the user has to specify two functions: *map* and *reduce* that are executed in parallel on multiple machines; the *map* part parses key/value pairs and passes them as input to the *reduce* function. Issues such as data splitting, task scheduling and fault tolerance are dealt with by the MapReduce framework in a user-transparent manner.

Google introduced MapReduce [5] as a solution to the need to process datasets up to multiple terabytes in size on a daily basis. An open-source implementation of the MapReduce model proposed by Google, is provided within the *Hadoop* project [9], whose popularity rapidly increased over the past years.

The MapReduce paradigm has recently been adopted by the Cloud computing community as a support to those Cloud-based applications that are data-intensive. Cloud providers support MapReduce computations so as to take advantage of the huge processing and storage capabilities the Cloud holds, but at the same time, to provide the user with a clean and easy-to-use interface. There are several options for running MapReduce applications in Clouds: renting Cloud resources and deploying a cluster of virtualized Hadoop instances on top of them, using the MapReduce service some Clouds provide, or using MapReduce frameworks built on Cloud services. Amazon released *Elastic MapReduce* [21], a web service that enables users to easily and cost-effectively process large amounts of data. The service consists in a hosted Hadoop

A. Hameurlain et al. (Eds.): Globe 2012, LNCS 7450, pp. 37–48, 2012.

framework running on Amazon's Elastic Compute Cloud (EC2) [20]. Amazon's Simple Storage Service (S3) [22] serves as storage layer for Hadoop. *AzureMapReduce* [8] is an implementation of the MapReduce programming model, based on the infrastructure services the Azure Cloud [2] offers.

In this work, we provide a cost evaluation of running MapReduce applications in the Cloud, by looking into two aspects: the overhead implied by executing the job on the Cloud, compared to executing it on a Grid and the actual costs of renting Cloud resources. To evaluate these factors, we run 3 MapReduce applications with the Hadoop framework in two environments: first on clusters belonging to the Grid'5000 [12] grid platform, then in a Nimbus [15] Cloud deployed on the Grid'5000 testbed. We then consider the payment scheme used by Amazon for the rental of their Cloud resources and we compute the cost of using our Nimbus deployment for running MapReduce applications. Our goal is to assess and understand the overhead generated by the execution of a MapReduce computation in a Cloud. More specifically, we analyze the *virtualization overhead* and the associated trade-offs when moving MapReduce applications to the Cloud. One of the major aspects to consider when porting an application to the Cloud regards the potential benefits and gains, if any. Moreover, it is important to be able to assess if those benefits are worth the costs. Apart from the obvious advantages of the Cloud (huge processing and storage capabilities), there are some application-related issues to take into account when choosing the right environment for running that application. In most cases, there is a compromise to make between the *cost* of executing the application in the Cloud, and the gains expected to be obtained. Therefore, it is highly important to understand the requirements and features of MapReduce applications, in order to be able to tune the Cloud environment in an optimal manner, so thus the right balance between cost and performance is struck. The main advantage of using *virtualization* is that one can create a homogeneous environment comprising a substantial number of machines by using a considerably lesser number of physical machines. In this work, we consider various MapReduce applications, with the goal of assessing the impact of replacing the typical MapReduce execution environment, i.e. a physical cluster, with a virtualized one, i.e. Cloud resources.

In Section 2 we present in detail the MapReduce applications we chose to evaluate. Section 3 introduces the cost model employed for our evaluation; we also describe the types of costs we take into account for our measurements. Sections 4 and 5 are dedicated to the experimental evaluation. In Section 6 we analyze the experimental results, in order to extract methods of minimizing the cost of running MapReduce applications in the Cloud. Section 7 briefly presents some of the on-going works on evaluating the cost of various types of applications in the Cloud.

2 MapReduce Applications: Case Studies

In general, the MapReduce model is appropriate for a large class of data-intensive applications which need to perform computations on large amounts of data. This type of applications can be expressed as computations that are executed in parallel on a large number of nodes. Many data-intensive applications belonging to various domains (from Internet services and data mining to bioinformatics, astronomy etc.) can be modeled using the MapReduce paradigm. In this paper, we take as study case three typical MapReduce applications:

Distributed Grep. The *grep* application searches through a huge text given as input file, for a specific pattern and outputs the occurrences of that pattern. This application is a distributed job that is data-intensive only in the "map" part of the job, as the "map" function is the phase that does the actual processing of the input file and pattern matching. In contrast, the "reduce" phase simply aggregates the data produced by the "map" step, thus processing and generating far less data than the first phase of the computation.

Distributed Sort. The goal of this application is to sort key/value pairs from the input data. The "map" function parses key/value pairs according to the job's specified input format, and emits them as intermediate pairs. On the "reduce" side, the computation is trivial, as the intermediate key/value pairs are output as the final result. The *sort* application generates the same amount of data as the provided input, which accounts for an output of significant size. *Distributed sort* is both read-intensive, as the mappers parse key/value pairs and sort them by key, and write-intensive, in the "reduce" phase when the output is written to the distributed file system. Because of its both read and write-intensive nature, this application is used as the standard test for benchmarking MapReduce frameworks.

Pipeline MapReduce Applications. Many of the computations that fit the MapReduce model, cannot be expressed as a single MapReduce execution, but require a more complex design. These applications that consist of multiple MapReduce jobs chained into a long-running execution, are called *pipeline MapReduce applications*. Each stage in the pipeline is a MapReduce job (with 2 phases, "map" and "reduce") and the output data produced by one stage is fed as input data to the next stage in the pipeline. Usually, pipeline MapReduce applications are long-running tasks that generate large amounts of intermediate data (the data produced between stages).

3 Computational and Cost Model

For our cost-evaluation of running MapReduce applications in the Cloud, we chose the Amazon services as the basic model. Amazon's EC2 Infrastructure-as-a-Service (IaaS) Cloud is the most widely-used and feature-rich commercial Cloud. The storage system introduced by Amazon, S3 [22], proposes a simple access interface that has become the IaaS standard for Cloud data storage.

Amazon EC2 allows users to rent compute or storage resources, in order to run their own applications. Typically, users first choose the type of virtual machine (VM) that suits their needs (application requirements, budget, etc.) and then boot the VM on multiple Amazon resources, thus creating what is referred to as *instances* of that VM. Users are charged on a pay-per-use model that involves 3 types of costs:

- **Computational cost** (CPU cost), that accounts for the VM type, number of instances and runtime.
- **Data storage costs** involve S3 charges for persistently storing input/output data for the executed applications.

- **Data transfer charges** include costs for moving data into and out of the Cloud. Data transfers between instances are free of charge, as well as transfers between S3 and the rented EC2 VMs.

To evaluate the computational cost of our experiments, we consider the *High-CPU Medium Instance* Amazon image type, as it meets two requirements: first, it has similar features to the physical nodes we used when measuring the overhead of moving applications to the Cloud. Second, in [14], the authors show that the *High-CPU Medium Instance* is the most cost-effective Amazon VM type. This instance is charged $0.19 per hour in the EU Amazon region and it features 1.7 GB of memory, 2 virtual cores and 350 GB of local storage.

One important parameter that influences the cost-analysis of a Cloud application is the granularity at which the Cloud provider charges for resources. In the case of Amazon EC2, the rented instances are charged by the hour, assumption that may conceal the benefits of adding resources to improve the runtime performance, when the execution lasts less than one hour. To better characterize the costs associated with our experiments, we will assume per-second charges in our cost model, by dividing the hourly prices in Amazon EC2 by 3600.

As for the storage costs, we consider Amazon S3 charges for the EU region, i.e., $0.125 per GB for the first TB per month. Regarding the transfer costs, Amazon charges only for data transfers out of the Cloud, that is $0.12 per GB for data downloaded from the Cloud (download is free for less than 1 GB of output data).

4 Execution Environment

To analyze the performance and costs of MapReduce applications, we performed experiments on two different platforms. To evaluate the pure performance of MapReduce access patterns, we relied on Grid'5000 [12], a large scale experimental testbed that provides the typical execution environment for MapReduce. Then, we deployed the same MapReduce tests in an IaaS Cloud deployed on top of the Grid'5000 testbed. To this end, we used the Nimbus [1, 15] toolkit, an open-source implementation of an Infrastructure-as-a-Service Cloud interface-compatible with Amazon's EC2. As a MapReduce framework for running our computations, we chose Hadoop [9], the widely used open-source implementation of Google's MapReduce model.

Hadoop. The Hadoop project comprises a large variety of tools for distributed applications. The core of the Hadoop project consists of *Hadoop MapReduce*, an open-source implementation of the MapReduce paradigm and HDFS [10], its default storage backend. The architecture of the Hadoop framework is designed following the master-slave model, comprising a centralized *jobtracker* in charge of coordinating several *tasktrackers*. Usually, these entities are deployed on the same cluster nodes as HDFS. When a user submits a MapReduce job for execution, the framework splits the input data residing in HDFS, into equally-sized chunks. Hadoop divides each MapReduce job into a set of tasks. Each chunk of input is processed by a map task, executed by the tasktrackers. After all the map tasks have completed successfully, the tasktrackers execute the reduce function on the map output data. The final result is stored in the distributed file system acting as backend storage, (by default, HDFS).

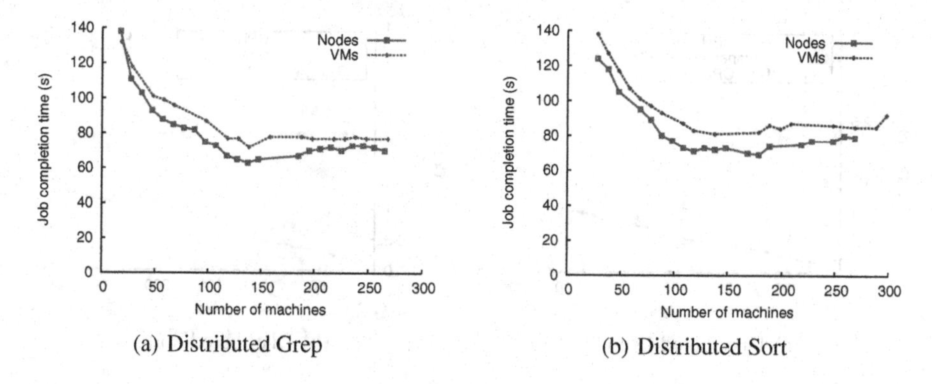

(a) Distributed Grep

(b) Distributed Sort

Fig. 1. Completion time for Hadoop applications

Experimental Platform. For both of our experimental environments (a physical cluster and the Cloud) we rely on Grid'5000 [12], a testbed widely spread across 10 sites in France and 1 in Luxembourg. The first set of experiments aims to run MapReduce applications in a typical cluster environment. For this setup, we selected the Grid'5000 cluster in Orsay, i.e., 275 nodes outfitted with dual-core x86 64 CPUs and 2 GB of RAM. Intra-cluster communication is done through a 1 Gbps Ethernet network. The second type of environment is Cloud-oriented: it was achieved by first deploying the Nimbus Cloud toolkit on top of physical nodes, and then by deploying VMs inside the obtained Nimbus Cloud. For these experiments, we used 130 physical nodes belonging to the Rennes site, and a VM type with features similar to the ones exhibited by the nodes from the first setup. Thus, we deployed VMs with 2 cores and 2 GB of RAM each.

In both setups, we created and deployed an execution environment for Hadoop comprising a dedicated node/VM for the jobtracker and another one for the namenode, while the rest of the nodes/VMs served as both datanodes and tasktrackers.

5 Results

5.1 Virtualization Overhead

In order to assess the virtualization overhead, we compare Hadoop's performance when running in the 2 experimental environments previously described (grid cluster versus Cloud). In both scenarios, we run the same tests that imply measuring the runtime of the *grep* and *sort* applications described in Section 2. A test consists in deploying both HDFS and the Hadoop MapReduce framework on a number of nodes/VMs and then running the 2 MapReduce applications. For each test, we fix the input data size to 18.8 GB residing in HDFS. The number of nodes/VMs on top of which Hadoop is deployed, ranges from 20 to 280. By executing the same workload and keeping the same setup when running Hadoop on the Grid and the in the Cloud, we manage to achieve a fair comparison between the 2 environments, and thus, to evaluate the virtualization overhead.

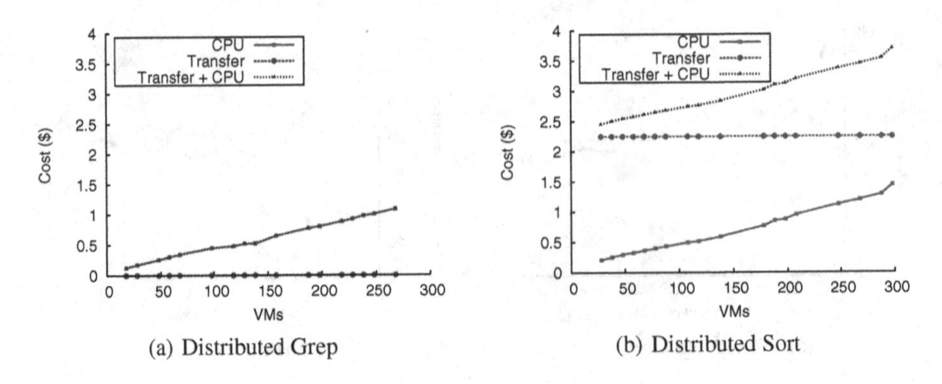

(a) Distributed Grep

(b) Distributed Sort

Fig. 2. Cost evaluation: computation (CPU) and data transfer

Figure 1 shows the completion time of *grep* and *sort* when Hadoop runs on physical and then on virtual machines. The results show that the job completion time decreases in both environments and for both applications, as the deployment platform increases, i.e. more nodes/VMs are used. However, this gain in performance stabilizes when a certain number of machines is reached. In our case, for the *grep* application (Figure 1(a)), the execution time does not improve further when using more than 150 machines for the deployment. The explanation for this behavior comes from the size of the input data and the scheduling policy of Hadoop: the jobtracker launches a mapper to process each chunk of input data. Considering the fact that we process 300 chunks of input data (18.8 GB) and that by default, Hadoop executes 2 mappers per node/VM, the optimal number for running this workload is 150 machines. After this point, the performance is almost the same. As *grep* is only read-intensive and produces very little output data, the completion time accounts mostly for computation time. For the *sort* application, the optimal number of machines involved in the deployment is around 100, since the completion time accounts not only for computation, but also for writing the output data. *Sort* generates the same amount of output data as the given input data, thus a considerably large amount of time is spent in writing the final result to HDFS.

These tests also help us assess the overhead of porting the *grep* and *sort* applications to the Cloud. As the results show, the virtualization overhead is negligible especially when considering the major benefit provided by the Cloud through virtualization. With a much smaller number of nodes, we managed to create inside the Cloud the same setup as the testing environment provided by a large number of physical machines in the Grid (double, in our case).

5.2 Cost Evaluation

Grep and Sort. In this section we evaluate the cost of running *grep* and *sort* applications in a Cloud environment. We focus on a typical scenario for Cloud services: the user submits a specific job to the service and transfers his input data into the Cloud; then he reserves a set of computational resources to execute the job and finally retrieves the obtained results. Such a scenario implies two of the types of costs detailed in Section 3,

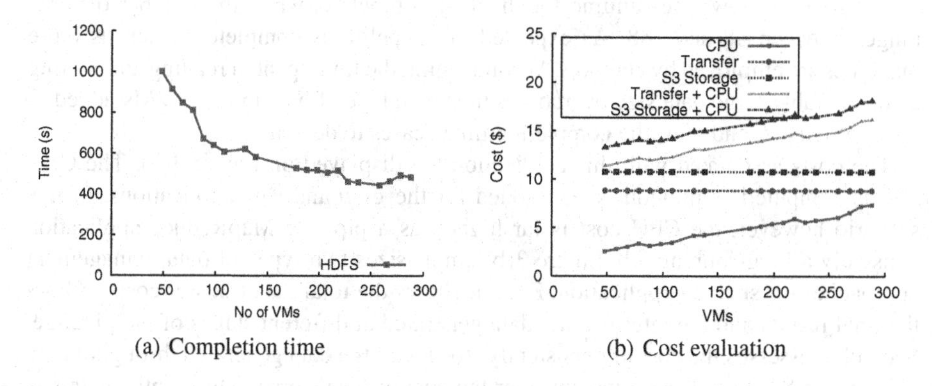

(a) Completion time (b) Cost evaluation

Fig. 3. Pipeline MapReduce application

namely the cost of computation and the cost of transferring the data into and from the Cloud.

Figure 2 shows the cost estimations of running each application as a function of the number of virtual machines that act as Hadoop nodes, for the same input size as in the previous experiment. The computational cost is defined as the number of VMs $\times$ the execution time $\times$ the cost of the VM instance per second. It increases as more VMs are provisioned, since the cost of deploying additional VMs outweighs the performance gain caused by the decreasing execution time.

The EC2 cost model assumes the data transfers into the Cloud are free of charge. The cost of the output data transfers does not vary with the number of deployed VMs, since the size of generated data is the same for each deployment setup. Both *grep* and *sort* process the same input data, that is 300 chunks of text files. As the *grep* application generates an negligible amount of output data, the cost of data transfers out of the Cloud is virtually zero, as shown in Figure 2(a). As a result, the cost of executing *grep* requests is only given by the cost of the rented virtual machines. In the case of the *sort* application, the amount of generated output data is equivalent to its input, that is 18.8 GB. Consequently, the data transfer cost has a significant impact on the total cost of the execution. As an example, when executing the job with the optimal number of VMs (i.e. 100 VMs in this scenario), the computation cost accounts only for 17% of the total cost.

Pipeline MapReduce Applications. To illustrate the characteristics and requirements of pipeline MapReduce applications, we developed a synthetic test which we then executed in a virtualized environment. Our synthetic application consists of 10 MapReduce jobs that are chained into a pipeline and executed with Hadoop. The computation performed by each job in the pipeline is trivial, as the "map " and "reduce" functions simply output key-value pairs. However, for our experiment, the computation itself is not relevant, as our goal is to execute a long-running pipeline application that generates large amounts of data. The input data consists of 200 chunks accounting for 12.5 GB. Each job in the pipeline parses key-value pairs from the input data and outputs 90% of them. This leads to a total amount of data to be stored in HDFS (input, intermediate and output data) of 85.8 GB.

Figure 3(a) shows the runtime for the 10-job pipeline, while the number of VMs ranges between 48 and 288. As expected, the pipeline is completed faster, as more machines are employed by Hadoop. At some point, the time spent in reading and writing a considerably large data size overcomes the advantage of having more VMs added to the deployment, and thus, the completion time ceases to decrease.

The costs associated with this application are displayed on Figure 3(b). The CPU cost is computed as previously mentioned for the *grep* and *sort* applications. In this scenario however, the CPU cost is far higher, as a pipeline MapReduce application is usually a long-running job. Figure 3(b) emphasizes two types of data management approaches for such an application. First, it shows the total data transfer costs of both the final results and the intermediate data generated at different stages of the pipeline. Second, it assesses the cost of persistently storing all the data generated throughout the pipeline, in S3. In this case, we consider the cost of data storage per month, as this is the default time interval for which charges are applied in Amazon S3. Both scenarios are plausible for such an application, which has two specific features. On the one hand, it generates a massive amount of data at each stage of the pipeline and consequently incurs a significant transfer or storage cost. On the other hand, each computation in the pipeline may generate significant results that can be interpreted by the user (after the data is transferred out of the Cloud) or further processed by other Cloud applications, justifying the need for persistent storage in the Cloud.

6 Discussion

The experiments we performed aimed at evaluating the impact of using a virtualized environment for executing MapReduce applications over the default scenario consisting of physical machines. As expected, porting an application to the Cloud incurs a performance overhead generated by the virtualization process. Our experiments have shown that the overhead ranges between 6% and 16%. This performance penalty is of minor impact and can be disregarded when taking into account the major benefit of virtualization: achieving a large, homogeneous environment using significantly less physical resources.

Achieving an optimal execution of MapReduce applications in the Cloud with respect to the cost of rented resources, is directly influenced by the workload of the application. As our experiments indicate, gains in performance can be achieved up to a point strongly related to the input data processed by the application. After this point, increasing the cost does not translate into additional gains. An in-depth analysis of the application's characteristics is a key point in tuning the Cloud environment.

The benchmarks we executed in a Cloud setup are representative for wider classes of MapReduce applications, each yielding different requirements. Our experiments set apart two Cloud use-cases that incur different types of costs. These use-cases are cost-effective for specific Map Reduce applications, while unprofitable for others. We distinguish two Cloud usage scenarios, based on the costs they involve. In both cases, we assume that Hadoop is deployed on Amazon EC2's VMs. A first scenario of running MapReduce in the Cloud consists of transferring the input data to the VMs, performing the MapReduce computation, and then obtaining the result by transferring the output data from the VMs. This scenario involves two types of costs: the transfer cost and the computational cost. Since data transfers to EC2 are free of charge, the transfer cost accounts only for copying the output data out of the Cloud. Thus, the total cost of the execution is computed as follows: $C_{total} = C_{transfer} + C_{CPU}$.

The second Cloud usage scenario entails also the storage cost. Users may decide to store the input data and/or the output data in S3, instead of transferring the data to the Cloud for each MapReduce computation. In this case, the total cost accounts for storing the data in S3 over a long period, and the computational cost of each processing executed on the stored data: $C_{total} = C_{storage} + C_{CPU}$.

Each of these two cost schemas are advantageous for specific MapReduce applications. Therefore, selecting the adequate schema is a crucial factor that impacts the obtained gains. An application that performs a single computation on a dataset will obviously employ the first scenario, since it does not need to store the data on long term. However, in a scenario where several queries are performed on the same dataset over a longer period of time, storing the data in S3 may turn out to be more cost-effective than transferring the data each time a query is executed.

To take an example of an application that would benefit from the second payment model, we consider CloudBurst [18], a MapReduce application that employs Hadoop for mapping next-generation sequence data to the human genome and other reference genomes. When running CloudBurst in the Cloud we have to consider several aspects. If users would want to lookup a certain DNA sequence (of a protein, for example) on a regular basis, they would have to execute the MapReduce application on the input data. Storing the data representing various types of human genomes in S3 and then running CloudBurst each time a user query is submitted, may prove to be more convenient than transferring the dataset for every query. In the first case, we compute the total cost as: $C_{total_1} = C_{storage} + n \times C_{CPU}$, where n represents the number of queries submitted by the users during the storing period. In the second case, $C_{total_2} = n \times (C_{transfer} + C_{CPU})$. If the data is stored on the Cloud, the application would benefit from low data access latencies (for input data) and the cost of these accesses would be zero. However, the storage cost can be significant; in order to be able to overcome the storage costs, the following condition must be satisfied: $C_{total_1} \leq C_{total_2}$, which leads to having $C_{storage} \leq n \times C_{transfer}$.

Thus, in order to benefit from storing the input data in S3, users would need to request at least a number of queries equal to $C_{storage}/C_{transfer}$. When storing the input data on the Cloud, an optimization that can be employed is to also store the output data generated by queries that are most frequently requested by users. In this case, $C_{total_3} = C_{storage_{input+output}} + C_{CPU}$. The computational cost for frequent queries accounts for running the queries only once.

7 Related Work

Several studies have investigated the performance of various Cloud platforms and the costs and benefits of running scientific applications in such environments. Most evaluations focused on the Amazon's EC2 Cloud, as it has become the most popular IaaS platform and has imposed its specific cost model to the Cloud computing community.

In [19] [7] [11], the authors explored the tradeoffs of running high-performance applications on EC2, showing that the Cloud environments introduce a significant overhead for parallel applications compared to local clusters. The cost of using HPC Cloud resources is discussed in [4], where the authors introduced a cost model for local resources and compared the computational cost of jobs against a Cloud environment. However, this work includes only a benchmark-based performance evaluation and no specific type of application is considered.

Several works have focused on loosely-coupled applications, such as [6] [14] [3], where the authors conducted a cost analysis of running scientific workflows in Cloud environments. They considered the performance penalties introduced by Cloud frameworks and evaluated computational and storage costs through simulations and experiments on EC2 and a local HPC cluster. More in-depth studies have investigated data storage in Clouds, evaluating the Amazon S3 service through data-intensive benchmarks [17]. Moreover, [13] evaluated several file systems as Cloud storage backends for workflow applications, emphasizing running times and costs for each backend. The work in [16] conducted a comparative evaluation of Cloud platforms against Desktop Grids. They examined performance and cost issues for specific volunteer computing applications and discuss hybrid approaches designed to improve cost effectiveness. In [8], the authors introduced the AzureMapReduce platform and conducted a performance comparison of several commercial MapReduce implementations in Cloud environments. The analysis included scalability tests and cost estimations on two MapReduce applications. In this paper however, our goal is to assess the specific requirements of various types of MapReduce applications in terms of Cloud processing resources and storage solutions, so as to optimize the execution cost. Moreover, we conduct our experiments on an open-source Cloud framework and we show it can sustain large MapReduce jobs through scalability tests.

8 Conclusions

This paper assesses the challenges posed by the execution of MapReduce applications on Cloud infrastructures instead of plain typical clusters. We evaluate the performance delivered to the users by measuring the completion time of 2 MapReduce applications, *grep* and *sort*, executed on the Hadoop framework in two different settings: first, we deploy Hadoop on physical nodes in Grid'5000; then we repeat the experiments using VMs provisioned by a Nimbus Cloud. Furthermore, we evaluate the computational and data management costs for running these applications in the Cloud, by considering Amazon services' charges as reference costs.

We also examine a special class of MapReduce computations, *pipeline applications* and we compute the costs of executing such an application in the Cloud and storing all the data it generates. Finally, we perform an overhead analysis of running MapReduce applications in Cloud environments and we discuss two Cloud data-management solutions suitable for such applications, as well as their advantages and drawbacks.

Acknowledgments. Experiments presented in this paper were carried out using the Grid'5000 experimental testbed, being developed under the INRIA ALADDIN development action with support from CNRS, RENATER and several Universities as well as other funding bodies (see `http://www.grid5000.org/`).

References

1. The Nimbus project, `http://www.nimbusproject.org/`
2. The Windows Azure Platform, `http://www.microsoft.com/windowsazure/`
3. Berriman, G.B., Juve, G., Deelman, E., et al.: The application of cloud computing to astronomy: A study of cost and performance. In: 2010 IEEE International Conference on e-Science Workshops, pp. 1–7 (2010)
4. Carlyle, A.G., Harrell, S.L., Smith, P.M.: Cost-effective HPC: The community or the cloud? In: The 2010 IEEE International Conference on Cloud Computing Technology and Science (CloudCom), pp. 169–176 (2010)
5. Dean, J., Ghemawat, S.: MapReduce: simplified data processing on large clusters. Communications of the ACM 51(1), 107–113 (2008)
6. Deelman, E., Singh, G., Livny, M., et al.: The cost of doing science on the cloud: the Montage example. In: Supercomputing 2008, Piscataway, NJ, USA, pp. 1–50. IEEE Press (2008)
7. Evangelinos, C., Hill, C.N.: Cloud Computing for parallel Scientific HPC Applications: Feasibility of Running Coupled Atmosphere-Ocean Climate Models on Amazon's EC2. Cloud Computing and Its Applications (October 2008)
8. Gunarathne, T., Wu, T.-L., Qiu, J., Fox, G.: Mapreduce in the clouds for science. In: Second International Conference on Cloud Computing Technology and Science (CloudCom), pp. 565–572 (2010)
9. The Apache Hadoop Project, `http://www.hadoop.org`
10. HDFS. The Hadoop Distributed File System, `http://hadoop.apache.org/common/docs/r0.20.1/hdfs_design.html`
11. Hill, Z., Humphrey, M.: A quantitative analysis of high performance computing with Amazon's EC2 infrastructure: The death of the local cluster? In: 2009 10th IEEE/ACM International Conference on Grid Computing, pp. 26–33 (October 2009)
12. Jégou, Y., Lantéri, S., Leduc, M., et al.: Grid'5000: a large scale and highly reconfigurable experimental Grid testbed. International Journal of High Performance Computing Applications 20(4), 481–494 (2006)
13. Juve, G., Deelman, E., Vahi, K., et al.: Data Sharing Options for Scientific Workflows on Amazon EC2. In: Supercomputing 2010, pp. 1–9. IEEE Computer Society, Washington, DC (2010)
14. Juve, G., Deelman, E., Vahi, K., et al.: Scientific Workflow Applications on Amazon EC2. In: 2009 5th IEEE International Conference on EScience Workshops, pp. 59–66 (2010)
15. Keahey, K., Figueiredo, R., Fortes, J., et al.: Science Clouds: Early experiences in cloud computing for scientific applications. In: Cloud Computing and Its Application 2008 (CCA 2008), Chicago (October 2008)
16. Kondo, D., Javadi, B., Malecot, P., et al.: Cost-benefit analysis of cloud computing versus desktop grids. In: IEEE International Symposium on Parallel Distributed Processing, pp. 1–12 (May 2009)
17. Palankar, M.R., Iamnitchi, A., Ripeanu, M., et al.: Amazon S3 for science grids: a viable solution? In: Proceedings of the 2008 International Workshop on Data-Aware Distributed Computing, pp. 55–64 (2008)

18. Schatz, M.C.: CloudBurst: highly sensitive read mapping with MapReduce. Bioinformatics 25(11), 1363–1369 (2009)
19. Walker, E.: Benchmarking Amazon EC2 for high-performance scientific computing. LOGIN 33(5), 18–23 (2008)
20. Amazon Elastic Compute Cloud (EC2), http://aws.amazon.com/ec2/
21. Amazon Elastic MapReduce, http://aws.amazon.com/elasticmapreduce/
22. Amazon Simple Storage Service (S3), http://aws.amazon.com/s3/

Scalability Issues in Designing and Implementing Semantic Provenance Management Systems

Mohamed Amin Sakka[1,2] and Bruno Defude[2]

[1] Novapost, Novapost R&D, 32, Rue de Paradis 75010 Paris-France
amin.sakka@novapost.fr
[2] TELECOM SudParis, CNRS UMR Samovar, 9,
Rue Charles Fourier 91011 Evry cedex-France
{mohamed_amin.sakka,bruno.defude}@it-sudparis.eu

Abstract. Provenance is a key metadata for assessing electronic documents trustworthiness. Most of the applications exchanging and processing documents on the web or in the cloud become provenance aware and provide heterogeneous, decentralized and not interoperable provenance data. A new type of system emerges, called provenance management system (or PMS). These systems offer a unified way to model, collect and query provenance data from various applications.

This work presents such a system based on semantic web technologies and focuses on scalability issues. In fact, modern infrastructure such as cloud can produce huge volume of provenance data and scalability becomes a major issue.

We describe here an implementation of our PMS based on an NoSQL DBMS coupled with the map-reduce parallel model and present different experimentations illustrating how it scales linearly depending on the size of the processed logs.

1 Introduction and Motivations

1.1 Introduction

The provenance of a piece of data is the process that lead to that piece of data. It is metadata recording the ultimate derivation and passage of an item through its various owners. Provenance can be used in different application areas and for different purposes like reliability, quality, re-usability, justification, audit, migration. It provides also an evidence of authenticity, integrity and information quality.

With the maturation of service oriented technologies and cloud computing, more and more data are exchanged electronically and dematerialization becomes one of the key concepts to cost reduction and efficiency improvement. Different services used for creating, processing, exchanging and archiving documents become available and accessible. In this context, knowing the provenance of data becomes extremely important.

In general, provenance data is sparse through different log files. These logs are structured using different formats (database, text, XML ...), use a suitable

A. Hameurlain et al. (Eds.): Globe 2012, LNCS 7450, pp. 49–61, 2012.

query language (SQL, XQuery ...), are semantically heterogeneous and can be distributed across servers. For example, an application can produce provenance data in a generic log file like a http log (generated by a http server) and in a dedicated (application-oriented) one (generated by application). These provenance sources need to be semantically correlated in order to produce an integrated view of provenance.

For these reasons, provenance management systems (PMSs) addressing the following points are needed:

- support of syntactic and semantic heterogeneity of provenance sources,
- support of rich domain models allowing to construct high level representation of provenance,
- support of semantic correlation between different domain models,
- support of high level semantic query languages,
- scalability.

In our previous work [1], we have designed a PMS based on semantic web models allowing to respect the modelling requirements. In this paper, we address the scalability issues by experimenting a NoSQL based implementation of our proposal on real datasets coming from Novapost[1]. Novapost is a French company specialized on providing collection, distribution and legal archiving services for electronic documents. Novapost must ensure document life cycle traceability as well as the compliance of all processing with the relative regulations. We were consulted by Novapost to define a provenance management approach corresponding to their needs and to design their PMS architecture.

The main contributions of this work are: (1) the proposition of a provenance framework for logs integration and enrichment based on semantic web standards and tools, (2) a logical architecture of our PMS and (3) an implementation of a PMS based on a NoSQL DBMS scaling up linearly with a limited decrease in expressiveness.

The rest of the paper is organized as follows. In section 2 we start by detailing the components of the proposed framework based on semantic web technologies. Section 3 presents the logical architecture of the provenance management system. Section 4 presents an implementation of the proposed architecture using a NoSQL DBMS and a set of experimentations. In these experiments, we evaluate provenance data load time and the average queries execution time for different datasets. We discuss related work in section 5. Finally, we conclude and describe future work in section 6.

2 A Provenance Management Framework Based on Semantic Web Technologies

We aim to define a provenance framework offering principally two degrees of liberty: (i) allowing to provide a unified view of heterogeneous provenance datasources using a common domain model and (ii) providing different views on a single provenance datasource for different end users.

[1] www.novapost.fr

To address these issues and facilitate provenance management, we design a provenance framework based on semantic web technologies. It offers an application independent way to collect and query provenance data.

Within our framework, the provenance of any document is extracted from log files (Cf. Fig 1 part A) and is represented using a semantic model. The basis of this model is the minimal domain model (or MDM for short, see Figure 1 part B). It contains only provenance informations without any reference to a particular domain (what is the action performed, on which document, by whom and when). It is based on the Open Provenance Model (OPM)[2]. This model ensures the genericity of our approach and the support of basic semantic heterogeneity between logs. The MDM can be specialized to define different domain models and describe their domain constraints (Cf. Fig 1 part B). These models allow to support semantic heterogeneity and to give different views of the same log. OWL[2] can be used to describe domain models (including MDM) thanks to its expressiveness and its associated tools such as graphical editors. But we will see in next sections that domain models can also be described as documents in a NoSQL document oriented DBMS.

Most of real applications will not use these models as their native log formats. Consequently, provenance providers should define an import function taking as input a provenance source and the corresponding domain model (called import domain model) and producing an instantiated domain model (Cf. Fig 1 part C). This model will be queried by end-users with the suitable query language for the used storage technology. Between the instantiated import domain model and the instantiated minimal domain model, a lattice of instantiated domain models can be created by means of a *generalization* relationship. This relationship permits to automatically generate a new, more abstract instantiated domain model that can be also queried. The whole concepts and details of our approach were presented in [1], its formalisation and details are out of scope of this paper.

3 Logical Architecture of the Provenance Management System

We present in this section the logical architecture of our semantic PMS. This architecture is summarized in Figure 2.

Our PMS is based on three main components offering administration, management and querying functionalities for provenance administrators and users. The PMS can be deployed in a unique location (a kind of provenance datawarehouse) or in multiple ones (for example one per service or per cloud provider). In the second case, the multiple PMSs have to be federated to allow end-users a unified access to provenance data. We are currently working on the design of the federated architecture, but this is out of scope of this article. The modules of the PMS are: 1) a provenance management module offering the functionalities of importing provenance data, managing provenance sources as well as domain models

[2] Web Ontology Language: `http://www.w3.org/TR/owl-features`

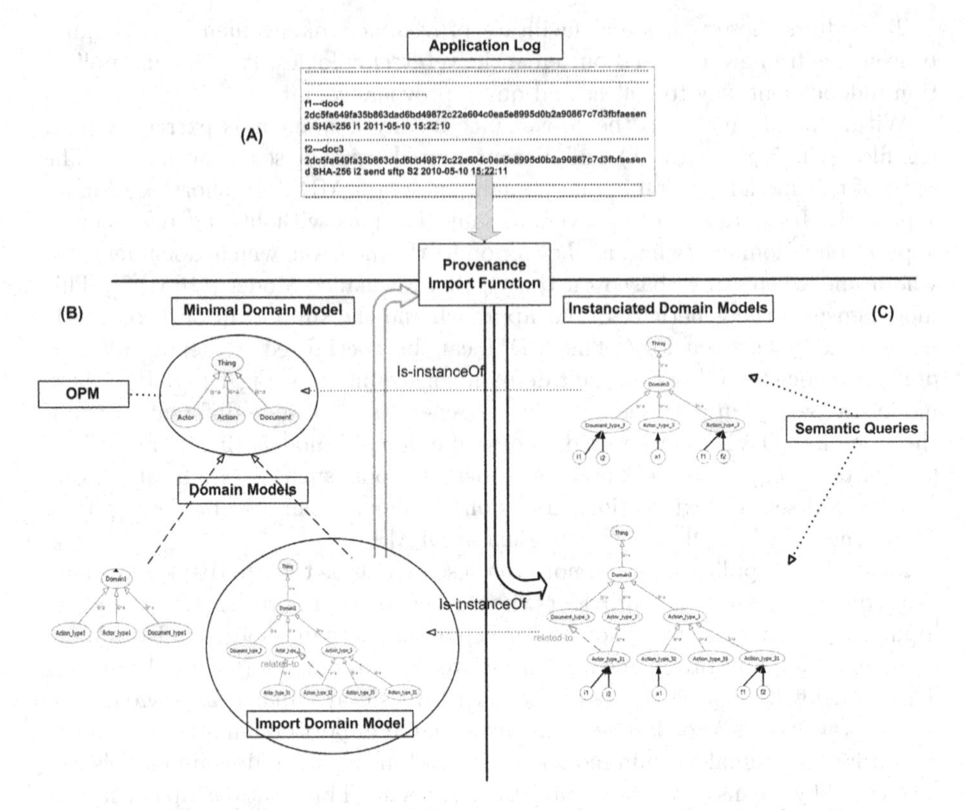

Fig. 1. Components of the semantic provenance framework

(DMs), instantiated domain models (IDMs) and fusion of IDMs, 2) provenance querying module and 3) provenance storage module.

The first module, that of provenance management provides a first functionality allowing to import provenance data from raw format into the provenance store via the import function. This module offers also to the administrator the capability to manage provenance sources like adding new sources, dropping an existent one or defining a new import function. This module provides also functionalities for domain models management like adding a new domain model by specializing the MDM or another DM and deleting an existing DM. The provenance management module offers also IDMs generalization allowing to generate different views of the same provenance data. With this generalization, we generate automatically new IDMs corresponding to different domain models. The last functionality of this module is fusion management. This fusion is performed when provenance comes from several sources. It aims to aggregate multiple sources of information, to facilitate further querying and combine these sources to generate new information that was not visible when the sources were separated. Thus, it is responsible on checking the semantic compatibility of the input IDMs that will be merged, checking any inconsistencies between IDMs instances and resolving them when they exist (see our work in [1] for more details).

The second module is that of query processing which is responsible on answering end-users queries. This module takes as input a provenance query and returns a complete provenance chain. The storage module is used for long term provenance storage. Physically, this module can be implemented by any storage technology. It can be a relational DBMS, an RDF triple store, a NoSQL DBMS ... But of course, the choice of a specific storage system will constraint the supported query language. The provenance store module can offer two APIs: one corresponding to the use of import functions and the other one to the direct storage of application logs if they are natively formatted with our models.

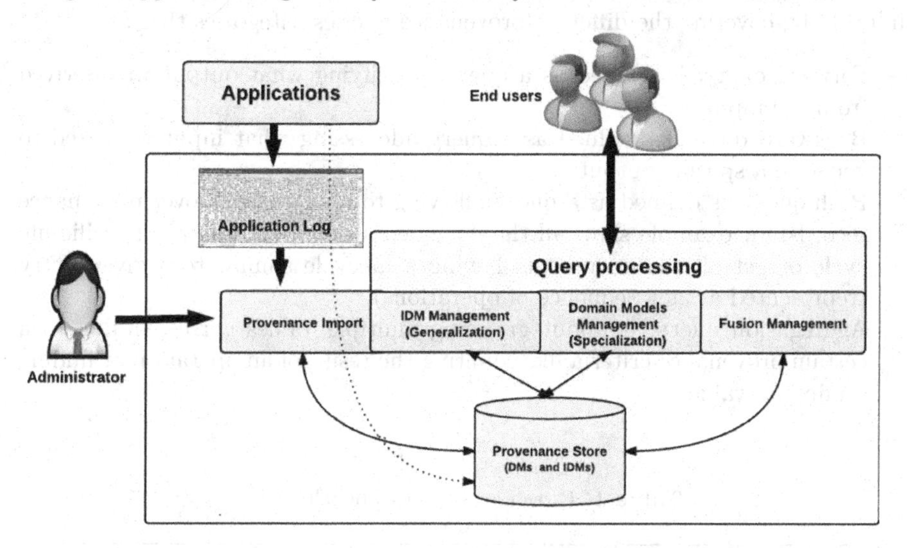

Fig. 2. Logical architecture of the Provenance Management System

4 Implementation with CouchDB, Experimentations and Scalability

In this part, we validate our approach by using real dataset from the company Novapost. The experiments we performed aim to validate the feasibility of the proposed approach, to test its performance and scalability.

We propose an implementation of our framework based on a NoSQL DBMS (CouchDB[3]), coupled with a map-reduce based query interface. Experiments were performed on Novapost's datasets coming from two provenance sources: the log of Tomcat server having 850 MB size and the log of the Apache http server having about 250 MB size. These two provenance sources contain traces of processing operations, distribution and archiving of documents of four months for about twenty customers and a total of 1M pdf documents, mainly payslips. This is about 2.5 million log lines that must be analysed and integrated into the PMS's storage module. This module was implemented based on a NoSQL DBMS.

[3] `http://couchdb.apache.org`

All experiments were conducted on a DELL workstation having the following configuration: Intel Xeon W3530 2.8GHz CPU; 12GB DDR2 667 RAM; 220GB (10,000 rpm) SATA2 hard disks and Fedora 13 (Goddard) 64-bit as operating system. We have installed (CouchDB v1.1.0) and BigCouch (v0.4a) on that workstation.

4.1 Provenance Queries Categories

Our experiments are based on a queries benchmark we have designed (presented in Table 1), covering the different provenance queries categories that are:

- Forward query: is defined as a query identifying what output was derived from the input.
- Backward query: is defined as a query addressing what input was used to generate a specific output.
- Path query: is defined as a query allowing to query a path over provenance records, for example show all the documents that have follow a specific life cycle or get all the documents having a lifecycle similar to a given entry (represented as as a sequence of operations).
- Aggregation query: is about grouping multiple values corresponding to a certain provenance criteria like counting the result of an operation or finding a minimal value.

Table 1. Provenance queries benchmark

	Provenance Query
Query 1 - Backward query	What are the documents processed by Novapost's archiving module ?
Query 2 - Forward query	What happens to the document "doc-02-10-C" after the split operation were performed on it ?
Query 3 - Aggregation query	What is the number of documents generated from the document "doc02-10-C" and that are archived in users electronic vault within "CDC" third party ?
Query 4 - Path query	What is the full provenance of the document "doc-02-2010-s2" based on the Tomcat server log ?
Query 5 - Path query	What is the full provenance of the document "doc-02-10-s2"?
Query 6 - Path query	What is the full provenance of the document "doc-02-2010-s2" from an auditor perspective ?

4.2 Representing and Querying Provenance in CouchDB

Before using a NoSQL DBMS, we have tested Sesame[4]. The choice of an RDF store was quite natural for implementing a PMS based on semantic web technologies. Sesame offers a rich query language (SPARQL) but our experiments performed with the same benchmark queries and datasets have shown that Sesame will not scale for large provenance data volumes.

[4] www.openrdf.org

Figure 3 illustrates the response time of Sesame and CouchDB for queries Q3 and Q6. It shows that Sesame is an order of magnitude slower than CouchDB for Q3 and furthermore does not scale up linearly especially for Q6. For these reasons we have decided to use a NoSQL DBMS rather than a RDF store.

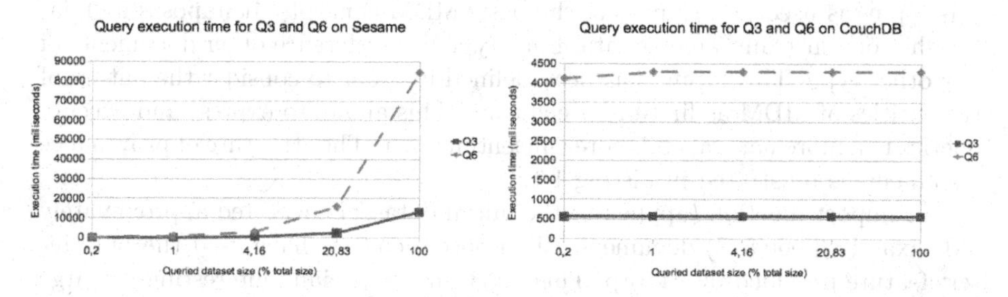

Fig. 3. Query response time in CouchDB for Q3 and Q6 on Sesame (left) and CouchDB (right)

NoSQL DBMS is a new word designating different types of non relational DBMS, that is key-value DBMS, document DBMS and graph DBMS. They have gained a lot of popularity these days with their use by big web players such as Google (BigTable) or Facebook (Cassandra) for example. Many of these NoSQL DBMS are coupled with a data parallel model called map/reduce [3] initially designed by Google and popularize by its open-source implementation Hadoop. Map-reduce allows to evaluate complex processing on large data sets using parallelism with a simple paradigm (map which maps data to nodes and reduce which aggregates the result produced by the nodes). Even if there are many discussions on the effectiveness of NoSQL DBMS related to parallel DBMS [4], NoSQL DBMS seem to provide an efficient and cheap solution to handle queries on large data sets [5]. In our context, we believe that the use of NoSQL database for large scale provenance management is interesting for the following reasons:

1. Provenance data structure is flexible. This corresponds well to the schema free characteristic of NoSQL databases,
2. Provenance data does need transactional guarantees when integrated,
3. Provenance data is static, never updated and is used in an append only mode,
4. Provenance data is characterized by its large volumes what requires efficient query techniques,
5. Some NoSQL databases provide querying techniques based on map/reduce what allows efficient selection and aggregation of data.

We have chosen CouchDB which is a document oriented NoSQL database written in Erlang. JavaScript (or Erlang) functions select and aggregate documents and representations of them in a MapReduce manner to build views of the database which also get indexed. Cluster of CouchDB nodes can also be used to ensure horizontal scalability.

We have developed an import function in Java. This function uses Ektrop[5] which is a CouchDB Java client facilitating the interaction with CouchDB and providing insert, bulk insert and querying functionalities.

We have chosen to represent instances of IDM as three distinct document types in CouchDB: **ActorDocument**, **ActionDocument** and **Document**. Indeed, this typing is used just to project the basic MDM concepts. It imposes no relationship or constraint among data. Each type may reference other documents of any other type. This typing has been defined in order to consider the nature of the entities of MDM as first-order elements. This allows to express and execute queries in a more natural and more efficient manner. The structure of provenance documents is illustrated in Listing 1.1.

Our import function (applied to the initial dataset) generated approximately 3M (exactly 3.000.517) documents. To insert them, we have used the bulk insert feature provided by Ektorp. This function is especially interesting for large datasets.

Listing 1.1. JSON representation of provenance documents stored in CouchDB

```
Document:
{
   "_id": "001e38d82a5818ea60dee3cef1032d1c",
   "_rev": "1-2dee8a50897071be3140dabac4290233",
   "type": "Document",
   "status": "sent",
   "ids": [ "cegid:744438","nova:paie\_11-2011-custCE4"
   ],
   "path":[''output,split,744439,28/11/2011-21:32:55'' ],
   "sha256": "
        D12AB8F6200BD0AAE1B1F5B9B5317F8F4113B2B9C015B3734045FA463B5A6D0D",
   "identifier": "744438"
}
```

Once our provenance data created and loaded into CouchDB, we should define the views permitting to query it. Views are the primary tool used for querying CouchDB databases. They can be defined in JavaScript (although there are other query servers available). Two different kinds of views exist in CouchDB: permanent and temporary views. Temporary views are not stored in the database, but rather executed on demand. This kind of query is very expensive to compute each time they get called and they get increasingly slower the more data you have in the database. Permanent views are stored inside special documents called design documents. This document contains different views over the database. Each view is identified by a unique name inside the design document and should define a map function and optionally a reduce function:

- The map function looks at all documents in CouchDB separately one after the other and creates a map result. The map result is an ordered list of key/value pairs. Both key and value can be specified by the user writing the map function. A map function may call the built-in emit(key, value) function 0 to N times per document, creating a row in the map result per invocation.

[5] www.ektorp.org

- The reduce function is similar to aggregate functions in SQL, it computes a value over multiple documents. This function operates on the output of the map function and returns a value.

The following listing (Cf. Listing 1.2) presents a view sample among the views we have defined to query CouchDB. Even if these javascript expressions are not too complex to define and to understand, it is clearly not as declarative and readable as SQL like queries.

Listing 1.2. Q3: Aggregation query

```
''documents_archived_number": {
          "map":  "function(doc) {if((doc.type=='Document') && (doc.status)
              && (doc.archiver) &&(doc.identifier))
                      emit([doc.identifier ,doc.status ,doc.archiver],
                      doc._id)}",
          "reduce":  ''_count"
    }
```

4.3 Experiments with CouchDB

Provenance Data Loading and Querying. We have used the bulk insert function provided by Ektrop to load our provenance data. Loading time illustrated in Table 2 -(a) shows that CouchDB is able to load large data volumes in a linear time.

Table 2 -(b) shows the time required to compute the materialised views defined in design documents for two different datasets. This time is linear and depends on the volume of data and the view type (simple or parametrised view). Adding new data to already indexed data requires additional time to re-compute the materialised view. This time depends on the added data volume.

Analysing Table 3, we notice that CouchDB presents almost constant time for different data size. This is also true for aggregation queries (Q3) and path queries (Q4, Q5 and Q6).

Table 2. Provenance data load time into CouchDB in seconds (a) and views computing time in seconds (b)

(a)

Log lines number (K)	Loading time
50	20.89
500	219
1000	427
2500	1044

(b)

Documents number (M)	1	6
Computing views time	175	363
+ 10 % new data	26	41

Our experiments on data loading and querying in CouchDB show that:

- CouchDB needs few seconds (or minutes) to compute the materialised views defined in design documents. Once the materialised views computed, we pay only the access time to data.

Table 3. Arithmetic mean of queries execution time for the different datasets in CouchDB (in milliseconds)

	Q1	Q2	Q3	Q4	Q5	Q6
250K	58	48	565	4227	4231	4227
1M	61	49	566	4325	4328	4282
3M	63	51	566	4331	4282	4285

- forward and backward queries are quite efficient while aggregation queries are an order of magnitude slower,
- path queries are not so efficient but their execution time remains reasonable,
- the execution time of all queries is independent from the dataset size. Map/reduce queries are efficient on small and large dataset,
- view definition (defining the map/reduce function for each query) is complicated for complex queries like path queries. To answer this category of queries, we have added a field called "path" where we provide the action, its type, its input and its timestamp. This field is created and enriched when loading the provenance data into CouchDB. However, this type of queries will be facilitated in the feature by the integration of chained map-reduce views in CouchDB.

Provenance Data Scalability. We aim to build a scalable semantic provenance management system. We have succeed to insert 15M documents into a single CouchDB node but we encountered memory problems while computing the materialized views. To test larger datasets, we have created a CouchDB Cluster and have distributed the database storage and indexation. For that, we have used BigCouch[6] which is an open source system provided by Cloudant permitting to use CouchDB as a cluster.

To analyse the response time of large data volumes, we have configured a three machines cluster (having the same hardware configuration of that mentioned in 4. We create a distributed database on which we have integrated respectively 15 (dataset D1) and 30 (dataset D2) millions provenance documents (i.e CouchDB documents). The dataset has been fragmented horizontally, i.e each node receives 1/3 of the provenance data. These experiments aim to test the impact of data size on provenance queries. To do that, we run our provenance queries and we measure the average runtime over 10 executions for datasets D1 and D2. The results are illustrated in Table 4-(a) and shows that the execution time is still independent of the size of the datasets due to the distribution of the query load on the different servers.

The last part of our experiments is focused on path queries. These queries compute intermediate results in a recursive manner. We want to analyse the impact of the size of the intermediate results on these queries. We want to know if the response time of these queries is related only to the number of intermediate results and not the amount of data. For this, we perform tests on Q6

[6] `http://bigcouch.cloudant.com`

Table 4. Average queries runtime in BigCouch/CouchDB (milliseconds) (a) and Impact of the size of intermediate results on Q6

(a)

Query	Execution time for D1	Execution time for D2
Q1	60	63
Q2	51	53
Q3	558	560
Q4	4339	4340
Q5	4345	4347
Q6	4343	4346

(b)

Number of intermediate results	6	15	20
Execution time for D'1	4285	6829	9374
Execution time for D'2	4343	6835	9387
Execution time for D'3	4359	6829	9397

with 3 (dataset D'1), 15 (dataset D'2) and 30 (dataset D'3) million provenance documents respectively. For each dataset, the first test involves six intermediate results, the second 15 and third 20. The results in Table 4-(b) show that the size of the intermediate results has a significant impact on the execution time. An increase of 333% in size implies an increase of 215% in execution time.

Also, the tests performed on BigCouch/CouchDB and that are illustrated in table 4-(a) shows that it scales in a linear fashion for all provenance queries. So, if you want to store larger provenance data, you just have to add a new server to the cluster. Nevertheless, one can notice that the per node scalability is very poor since we have not succeed to store and query around 50 Gigabytes per node. This low efficiency seems related to a configuration problem on CouchDB and we have to investigate in this direction to be able to demonstrate that our system scales with high efficiency.

5 Related Works

Provenance challenges and opportunities has been the target of different research works[6,7]. Many approaches center on a "workflow engine perspective of any system" and consider that operations are orchestrated by that workflow engine. This vision ignores what is domain specific relevant (semantics, relationships). To address this issue, provenance modeling and management aspects were addressed in other works [8,9] and different frameworks were proposed [10]. Principally, they are based on one of the following management approaches:

- group provenance and business specificities in a unique and semantically rich model like W7[11],
- propose a minimal, semantically poor model that can be annotated and enriched like OPM[2], PrOM[12,13] and Hartig's model [14]. For this approach, the enrichment techniques of the minimal model are the keys. Such techniques should be flexible and offer easy collect and query possibilities.

Other research works have tackled design and implementation issues of PMSs for storing and querying provenance data. Groth et al [10] have proposed within the European project Provenance[7] a logical architecture for the implementation of a PMS. This architecture describes a provenance model structure based on p-assertions. Security, scalability and deployment aspects in industrial environments were considered too. However, this architecture does not consider the use of existent provenance sources and may by cumbersome to integrate.

Sahoo et al [13] have proposed the PrOM framework that includes a scalable provenance query engine supporting complex queries over very large RDF datasets. This query engine uses a new class of materialized views for query optimization. Zhao et al [15] have proposed an approach for querying provenance in distributed environments. This approach is based on a provenance index service keeping the mappings between the provenance repositories and the data artifacts. This approach considers just two categories of provenance queries. Chebotko et al [16] have proposed RDFPROV, a relational RDF store based on a set of mappings (schema mapping, data mapping, and SPARQL-to-SQL) to combine semantic web technologies advantages with the storage and querying power of RDBMs. However, RDFPROV do not address the issues about provenance loading and PMS scalability.

In our work, we have addressed different dimensions of provenance management within a global approach aiming to build a scalable semantic provenance management system. Unlike W7[11], we have proposed to handle provenance modeling using an OPM-like [2] vision based on a minimal and semantically poor model that can be enriched. We have proposed a logical architecture for PMSs and combines two type of technologies: semantic web for provenance modeling and correlation and NoSQL/Map-reduce for provenance storage and querying. Our work have also addressed provenance data loading and querying issues and have provided different interpretations about provenance queries expressiveness and scalability.

6 Conclusion

As data is shared across networks and exceeds traditional boundaries, provenance becomes more and more important. By knowing data provenance, we can judge its quality and measure its trustworthiness. In this paper, we present a provenance framework based on semantic web technologies to model, store and query heterogeneous and distributed provenance sources. it allows to describe provenance data with different domain models which are all a specialization of the minimal data model (ensuring interoperability).

This paper focuses on the design and the implementation of scalable semantic provenance management systems. A logical architecture of a provenance system has been designed and we have choose the use a NoSQL document-oriented database (CouchDB) as a storage technology. We have proposed a document storage structure on CouchDB that we have evaluated. The results show that CouchDB scales up linearly for the different queries. Also, it presents good performance for aggregation and path queries. However, per node efficiency is not very good and we have to tune CouchDB configuration to improve it.

[7] www.gridprovenance.org

As on-going works, we are working on the unified federation of PMSs, corresponding to different service or cloud providers. The idea here is to use a mediator-based architecture on top of PMSs. The mediator will be in charge of the correlation between provenance sources and of the query rewriting on the PSMs. Specific rewriting methods can be used because this mediator will have a minimal knowledge of all the instances and so can achieve better optimizations.

References

1. Sakka, M.A., Defude, B., Tellez, J.: A semantic framework for the management of enriched provenance logs. In: Proc. of the 26th AINA Conference. IEEE Computer Society (2012)
2. Moreau, L., Clifford, B., Freire, J., Futrelle, J., Gil, Y., Groth, P., Kwasnikowska, N., Miles, S., Missier, P., Myers, J., Plale, B., Simmhan, Y.L., Stephan, E., Bussche, J.V.: The open provenance model core specification (v1.1). In: FGCS (2010)
3. Dean, J., Ghemawat, S.: Mapreduce: a flexible data processing tool. Commun. ACM 53(1), 72–77 (2010)
4. Stonebraker, M., Abadi, D., DeWitt, D.J., Madden, S., Paulson, E., Pavlo, A., Rasin, A.: Mapreduce and parallel dbmss: friends or foes? Commun. ACM 53(1), 64–71 (2010)
5. Pavlo, A., Paulson, E., Rasin, A., Abadi, D.J., DeWitt, D.J., Madden, S., Stonebraker, M.: A comparison of approaches to large-scale data analysis. In: Proceedings of the 35th SIGMOD International Conference on Management of Data, SIGMOD 2009, pp. 165–178. ACM, New York (2009)
6. Kiran Kumar, M.R.: Foundations for Provenance-Aware Systems. PhD thesis, Harvard University (2010)
7. Davidson, S.B., Freire, J.: Provenance and scientific workflows: challenges and opportunities. In: Proceedings of ACM SIGMOD, pp. 1345–1350 (2008)
8. Simmhan, Y.L., Plale, B., Gannon, D.: A survey of data provenance in e-science. SIGMOD Rec. 34, 31–36 (2005)
9. Freire, J., Koop, D., Santos, E., Silva, C.T.: Provenance for computational tasks: A survey. Computing in Science and Engineering, 11–21 (2008)
10. Groth, P., Jiang, S., Miles, S., Munroe, S., Tan, V., Tsasakou, S., Moreau, L.: An architecture for provenance systems. Technical report (February 2006), http://eprints.ecs.soton.ac.uk/13196 (access on December 2011)
11. Sudha, R., Jun, L.: A new perspective on semantics of data provenance. In: The First International Workshop on Role of Semantic Web in Provenance Management, SWPM 2009 (2009)
12. Sahoo, S.S., Sheth, A., Henson, C.: Semantic provenance for escience: Managing the deluge of scientific data. IEEE Internet Computing 12, 46–54 (2008)
13. Sahoo, S.S., Barga, R., Sheth, A., Thirunarayan, K., Hitzler, P.: Prom: A semantic web framework for provenance management in science. Technical Report KNOESIS-TR-2009, Kno.e.sis Center (2009)
14. Hartig, O.: Provenance information in the web of data. In: Second Workshop on Linked Data on the Web, LDOW (2009)
15. Zhao, J., Simmhan, Y., Gomadam, K., Prasanna, V.K.: Querying provenance information in distributed environments. IJCA 18(3), 196–215 (2011)
16. Chebotko, A., Lu, S., Fei, X., Fotouhi, F.: Rdfprov: A relational rdf store for querying and managing scientific workflow provenance. Data Knowl. Eng., 836–865 (2010)

Performance Characteristics of Virtualized Platforms from Applications Perspective

Balwinder Sodhi and T.V. Prabhakar

Department of Computer Science and Engineering
IIT Kanpur, UP India 208016
{sodhi,tvp}@cse.iitk.ac.in

Abstract. Virtualization allows creating isolated computing environments, often called virtual machines (VM), which execute on a shared underlying physical hardware infrastructure. From a application designer's perspective, it is critical to understand performance characteristics of such virtualized platforms. This is a challenging task mainly due to: a) There are multiple types of virtualization platforms such as Virtual Machine Monitor (VMM) based and operating system (OS) based. b) Most existing studies are focussed on evaluating narrow low-level subsystems of such platforms.

Three main forms of virtualization platforms – bare metal VMM, hosted VMM and OS based – are examined to understand their performance characteristics, such as CPU and memory usage, cross VM interference, from different dimensions. These characteristics are examined at macro level from applications perspective by subjecting these platforms to different types of load mix. One of our findings shows that the impact of VM co-location on different performance indicators is strongly dependent on platform and workload types.

1 Introduction

Central idea of hardware virtualization is to provide isolated computing environments on a shared hardware infrastructure. The isolated computing environments are often called Virtual Machines (VM) or Containers etc. A VM is executed by the Virtual Machine Monitor (VMM), e.g. Xen [1], on a physical hardware shared with other VMs. Typically, the software applications that are deployed on a VM cannot tell (or worry about) whether they are running inside a VM instead of a physical host. VMM technologies have gained renewed interest [2] both in the traditional area of data centers as well as of personal computing environments. Operating system (OS) based isolated virtual environments have been popular in UNIX like OSes. Example are Linux LXC [3] and Solaris containers etc.

Improving utilization by computing resource consolidation is an important goal of virtualization, particularly in data centers. Virtualization is at the heart of disruptive modern paradigms like Cloud Computing. A VM is a basic artifact delivered to cloud users in Infrastructure as a Service (IaaS) cloud.

Most existing performance characterization studies of virtualization are focussed either on a particular type of virtual platform, or a sub-system (e.g. I/O)

A. Hameurlain et al. (Eds.): Globe 2012, LNCS 7450, pp. 62–74, 2012.

of it. Characterization of such platforms, from application software perspective, is often seen left out. Typically, in a virtualized platform the consumption of a physical resource can be expressed as sum of consumption caused by: *i)* the programs running in the VM, *ii)* virtualization software running its own instructions, but running them in response to the activity in VM and *iii)* virtualization software running its own instructions, but not caused by any activity in the VM. Part *ii* is often called the overhead caused by and chargeable to the VM.

In the presented work, performance characteristics of various platforms are examined from the perspective of applications executing on them. For an application deployed in a VM it means that the quantities of interest are performance indicators (e.g. CPU, memory usage etc.) as viewed/reported inside the VM. This can be seen as roughly proportional to portion *i* above. The reason for focussing on said perspective is that in the environments like IaaS Cloud, applications are restricted to examine and control only the VM it is hosted in. Therefore, a prior knowledge of performance characteristics can be utilized by application architect as well as cloud provider. For instance, cloud provider can shape VM deployment decisions based on the nature of application's workload. The application architect, on the other hand, can design for mitigating the impact of adverse VM co-location.

Main goal of the presented work is to answer questions like below:

- How do the following vary with different types of a) virtualization and b) workloads:
  - Impact of VM co-location
  - Performance and resource consumption
- Which platform is best suited for a given requirement:
  - From the performance and resource consumption perspective
  - From co-location perspective

Both VMM based as well as OS based virtualization platforms have been benchmarked.

2 Related Work

Most performance evaluations of virtualized platforms reported in literature are either focussed on a micro area, such as disk I/O, of a single virtualization platform or are focussed on a small subset of platforms at macro level. For instance, Ahmad et. al. [4] present the disk performance analysis in VMware ESX server virtual machines. The authors do performance characterization of virtualized and physical systems via several disk microbenchmarks on different storage systems. Similarly, Menon et. al. [5] have investigated performance overheads in Xen VMM platform. The authors leverage the Xenoprof tool to examine kernel subsystems for various overheads in an Xen system. There are other works reported in literature which deal with the performance characterization of different subsystems of specific virtualization platforms.

Macro level performance evaluation of two types of virtualization platforms have been carried out by Padala et. al. [6]. They evaluate performance, scalability and low-level system metrics like cache misses etc. of Xen and OpenVZ [7] against a regular non-virtual system. They have used an eBay like auction application – RUBiS [8] – for their experiments. RUBiS is modeled to "evaluate application design patterns and application servers performance scalability".

Hosted VMMs (e.g. VirtualBox [9]) have become increasingly popular for use in the personal computing and volunteered computing space. In our survey of current literature we found that the hosted VMMs have been routinely left out of the performance evaluations. To the best of our knowledge, none of the evaluations cover all three major types of virtualization platforms – from hosted and bare-metal VMMs to OS based – on mix of main load types: CPU bound and I/O bound. The above gaps have been addressed in the presented work. We examine the performance properties and cross-VM interference of the said virtual platforms for different deployment and load type configurations.

3 Experiment Environment Setup

Almost every practical computing task can be classified as either CPU intensive, or I/O intensive or a mix of CPU and I/O intensive. As such, for characterizing the performance on different virtualization platforms, workload types are categorized into these three types. Performance characteristics of virtualization platforms are examined in a multi-tenant environment, that is, in scenarios where multiple VMs deploy different combinations of above said workload types. For instance, the above said three workload types can be deployed on two co-located VMs in nine different arrangements ($[No.\ of\ workload\ types]^{[No.\ of\ VMs]} = 3^2 = 9$). Table-1 shows the dimensions of our performance analysis.

Table 1. Analysis dimensions

Dimension	Range
Virtualization Types	Bare-metal VMM (Xen), Hosted VMM (VirtualBox), OS based (LXC)
Workload Scenarios	All 9 arrangements of {CPU-intensive, I/O-intensive, Mix of CPU and I/O} on 2 VMs
Performance Indicators	CPU usage, Memory usage, Throughput

Fig. 1 shows the VM setup on the physical host. Out of these two co-located VMs one was subjected to constant requests load e.g. 20 concurrent users, while the second was subjected to an increasing load. Performance characteristics are measured with respect to non-virtual Linux based OS running on a physical host. Henceforth, this platform is referred to as native platform. The virtualization based computing platforms that were considered are:

1. Xen (Bare metal VMM) version 4.0
2. Oracle VirtualBox (Hosted VMM) version 4.1.6r74727
3. Linux LXC Containers (OS based virtualization) version 0.6.5

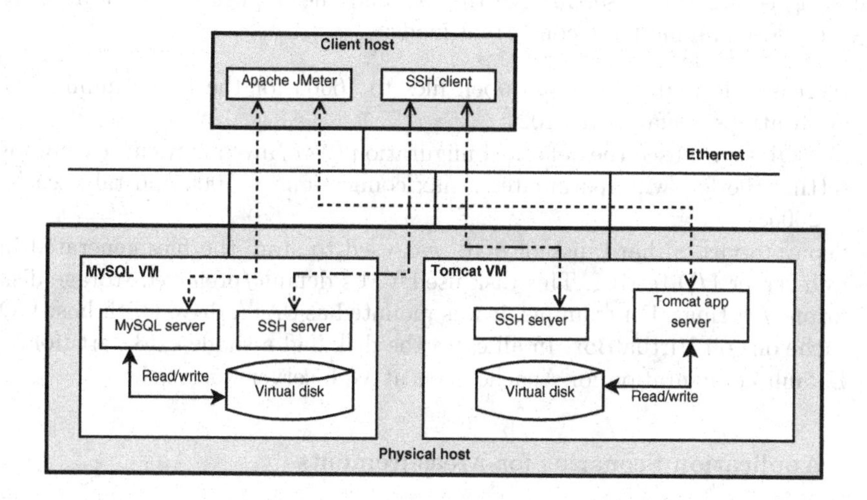

Fig. 1. Machines setup for testing

Table-2 shows the details of host environment that runs the guest VMs used for performance tests. Hardware assistance flags (VT-x, VT-d, VT-c) for virtualization were enabled.

Table 2. Host Platform Details

Item	Details
Operating system	Ubuntu 10.04.3 LTS Linux (kernel 2.6.32-39-generic)
CPU Model	Intel(R) Core(TM) i7 CPU 930 @ 2.80GHz
CPU Cores	4 (8 via hyper-threading)
CPU Cache size	L2 = 256KB and L3 = 8192KB
Address sizes	36 bits physical, 48 bits virtual
System memory	8GB SDRAM
Network Interface	10/100/1000 Mbps Ethernet on board
CPUs per VM	2
Memory per VM	2GB

Application software used for performance measurements are:

– MySQL database server version 5.1. It represents mix of CPU and IO intensive load type.

– Apache Tomcat [10] application server version 6.0.35 running on Java HotSpot(TM) 64-Bit Server VM. Java version 1.6.0_22. It is used to run CPU and IO intensive tasks.

Following configuration settings of the OS and the application software were used to allow running large concurrent loads:

– Increase the limit of allowed open files to 16000 for the user running test applications. Default was 1024.
– MySQL server used the default configuration (/etc/mysql/my.cnf) except for setting the following paramaters: max_connections = 1000 and table_cache = 2000.
– Separate virtual hard disk of 8GB was used to store the files generated in web server I/O testing. This disk used VM's default/preferred storage disk format/settings. For example, it was mounted as SATA drive (with host I/O cache on) on VirtualBox. In all cases the disk had a single ext3 partition.
– Default configuration for Apache Tomcat Web server was used.

3.1 Application Scenarios for Measurements

For performance characterization of virtualized platforms, following types of application software was used:

1. The Sakila Sample DB [11] from MySQL RDBMS.
2. Java applications software for:
   – CPU bound jobs
   – File I/O tasks
3. Web application server.

Fig. 1 depicts the architecture of our testbed. Performance characteristics of a VM were measured by subjecting it to a uniformly increasing requests load. Simultaneously, the co-located VM was subjected to a constant request load. Uniformly increasing load comprised of 10, 20, 30, 40, 60, 80 and 100 concurrent users each one performing a given set of operations for 20 iterations. Constant load comprised of 20 concurrent users firing a steady stream of requests. Delay between any two consecutive requests was kept between 1000 to 2000 ms.

In each of the scenarios, the following performance indicators were measured for the target process (i.e. mysqld and Tomcat JVM):

1. Variation in CPU consumption.
2. Variation in memory consumption.
3. Throughput for test operations executed by the process.

Apache JMeter [12] was used for generating concurrent user requests load. The measurements of CPU and memory usage was done via `top -b -p <pid>` command on Linux.

Rationale for Setup Choices. Our choice of application software was guided by the patterns of application software found in an average enterprise. Most such applications can be categorized into either predominantly CPU bound (e.g. image processing) or disk I/O bound (e.g. file servers), or an in-between mix (e.g. an database server). Hence, we have chosen applications that allow us to exercise these three types of loads. The number of concurrent users that were chosen are the minimum that would give a clear picture of performance variation pattern for platforms under consideration. It is important to note that our goal is not to benchmark the capabilities of software applications themselves under massive load scale, instead we are interested in identifying the performance variation and inter-VM interference characteristics of the virtualized platforms under a given deployment mix. In literature, we find main performance indicators of interest are often CPU and memory consumption and throughput. Hence we chose the said performance indicators for our experiments.

3.2 Load Profile for MySQL

In order to measure the performance characteristics of MySQL VM, the sample [11] DVD store database was used. The DVD database is modelled to represent a DVD rental business where users can search for movies DVDs and rent those. The sample DB comes pre-populated with different films, inventory and rental records. In benchmark tests, this DB is subjected to SQL operations that perform following tasks:

1. Rent a DVD. This involves a mix of SELECT, INSERT and UPDATE SQL statements being issued.
2. Return a rented DVD. This involves a mix of SELECT and UPDATEs.
3. Three different queries which perform SELECTs with varied number of joins on different tables.

MySQL Sakila sample database was chosen because: a) MySQL is a popular database server that drives a large majority of data driven applications in the field, b) Sakila sample database has a fairly complex schema (see [11]) consisting of 16 tables and 7 views. It is designed to represent and support complex on-line transaction processing capabilities.

3.3 Load Profile for Application Server

A Java web application deployed on Tomcat server was used for benchmarking performance characteristics. This web application consisted of a single Java Servlet which performed the following tasks for each request it received:

1. Create a file of given size containing random text.
2. Read and compress the above text file.
3. Delete the compressed and text files.
4. Return name and size of text and compressed files as response.

For creating the file with random text, a master text file with 10,000 different lines of text was used. Text lines from the master were randomly picked for creating the text files. Standard utilities (`java.util.zip.GZIPOutputStream` and `java.util.Random`) from Java SDK were used for picking random lines of text from master and for compressing the random text files. As is evident from the above description, Servlet served the purpose of generating CPU and disk I/O load for benchmark testing.

4 Measurement Results and Analysis

As mentioned in Section 1, we focussed on the performance characteristics of various platforms from the perspective of an application and its immediate hosting environment. Correlation coefficient (CC) and statistical mean values were calculated [13] from the collected data for different scenarios. CC is often used as an estimate of the correlation, i.e., linear dependence between two variable quantities. Its value lies in the range +1 and 1 inclusive. A +ve correlation between two variables indicates that as one variable increases, the second also tends to increase. A -ve correlation indicates the reverse, i.e., as one variable increases, the second tends to decrease. To obtain a better and conservative measure of the correlation between two variables x and y, we defined a Correlation Index (CI) as:

$$CI_{x,y} = CC_{x,y} \times min(\hat{\sigma}_x, \hat{\sigma}_y) \quad (1)$$

Here $\hat{\sigma}$ is the standard deviation (normalized to scale of 0-1) of the variable. The reason for using CI instead of CC is to eliminate those cases where, even though the two variables change together, the overall variation in the two variables differ significantly.

The CI for performance indicator values from the two VMs indicates the interference between co-located VMs for various workload deployment scenarios. Similarly, the mean values for different performance indicators allow us to compare different platforms for a given workload scenario and vice-versa. In all the plots (Fig. 2-6), x-axis shows the workload deployment scenarios for two co-located VMs. Each entry on x-axis is a hyphen separated tuple whose first element is the constant load workload type and second one is varying load workload type. For example, CPU-IO means that the VM subjected to constant load had CPU intensive workload and the VM subjected to varying load had an IO intensive workload.

4.1 Impact of VM Co-location

The benchmarking data from the experiments has been analysed to identify impact on performance isolation along the dimensions indicated in Table-1. Correlation information for different dimensions is shown in Fig. 2 and 3. We observe both +ve and -ve correlation across different dimensions. In the presented analysis, only CI values above 0.15 are considered to indicate any correlation.

Throughput. CI values for throughput across workload and platform types are plotted in Fig. 2. Prominent correlation is observed for three workload scenarios: CPU-IO, DB-IO and IO-IO. That is, when we have a varying IO intensive workload co-located with any of the workload types. The correlation is significant for native, LXC and VirtualBox. For both native and LXC cases, the correlation is -ve. For VirtualBox it is +ve. Interestingly, we do not observe any significant correlation for Xen for any type of workload co-location scenarios. Another interesting point to note is that VirtualBox does not have any -ve correlation in any scenario.

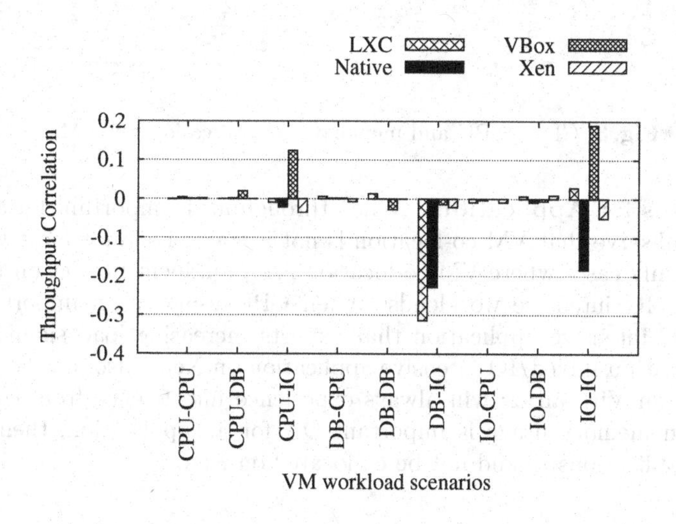

Fig. 2. CI for throughput of co-located VMs

CPU and Memory Consumption. CI values for CPU and memory consumption for co-located VMs running different types of workloads are plotted in Fig. 3. For CPU consumption, there is prominent correlation seen for CPU-IO, DB-CPU, IO-CPU and IO-IO workload types. For native platform we notice prominent -ve correlation for DB-CPU, IO-CPU and IO-IO workloads. LXC shows -ve correlation in DB-CPU and IO-CPU workloads. Only correlation shown by Xen is a +ve one for DB-CPU workload type. VirtualBox exhibits +ve correlation in just two workload scenarios: CPU-IO and IO-IO. For CPU consumption, no significant correlation is observed for CPU-CPU, CPU-DB, DB-DB, DB-IO and IO-DB workload scenarios.

For memory consumption, we observe correlation for all workload cases except CPU-IO, DB-CPU, DB-DB and DB-IO. Most prominent is exhibited by Xen for CPU-DB, IO-DB and IO-IO scenarios. VirtualBox shows correlation for CPU-CPU, CPU-DB, and IO-CPU cases. LXC showed correlation only in two cases: CPU-CPU and IO-CPU. Native platform shows correlation only for CPU-DB and IO-IO cases.

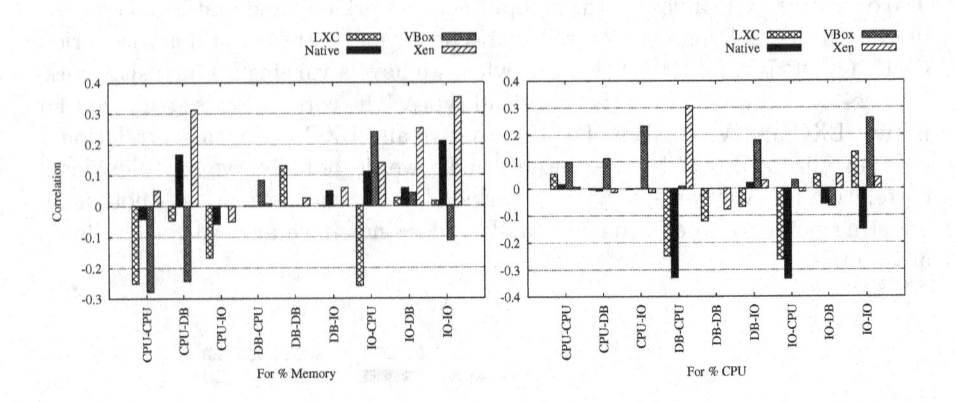

Fig. 3. CI for CPU and memory usage on co-located VMs

Implications for Applications. When throughput is important for an application, we observe that VM co-location is not a major issue for most workload scenarios. Only cases where VM co-location has prominent impact on throughput was for IO intensive workloads. When CPU usage is an important QA, then a CPU intensive application that expects increasing load should not be co-located with a mixed/IO intensive application on Xen. Also, an IO intensive application on VirtualBox will always experience interference from co-located VMs. When memory usage is important QA for an application, then two IO intensive applications should not be co-located on Xen.

4.2 Workload Performance on Different Platforms

Performance indicators measured on different platforms that run different types of workloads are shown in Fig. 4-6. The performance indicators are measures on both the co-located VMs – one subjected to a constant load and second to an increasing load.

Throughput. Throughput for different workload types running on various types of platforms is shown in Fig. 4 and 4. Throughput under constant load for different platforms is depicted in Fig. 4. Key observations from Fig. 4 are:

- Native platform always gives the best throughput for almost all the workload types.
- For CPU intensive workloads throughput on VirtualBox is about half on the rest.
- Throughput on LXC and Xen is almost same as native for most workload types. One scenario where LXC performance degrades is when it is running a mixed workload co-located with an IO intensive one.
- For mixed workload type, throughput on Native, Xen and VirtualBox is comparable.

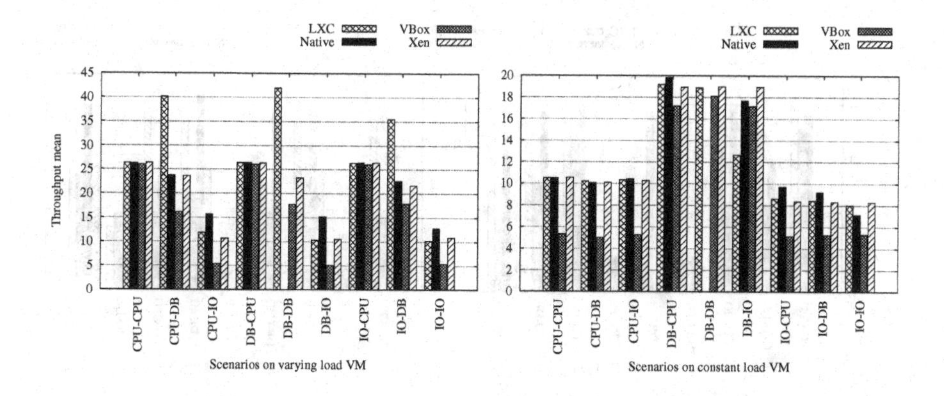

Fig. 4. Mean throughput for VMs (Requests/sec.)

- For IO intensive workloads, throughput on LXC, Native and Xen is comparable.

Fig. 4 shows throughput data for the VM subjected to increasing load. Following are the key observations from Fig. 4:

- For CPU intensive workloads, throughput is similar on all platforms.
- For IO intensive workloads, throughput is similar for LXC, Native and Xen. VirtualBox has lower throughput here.
- For a mixed workload, LXC gives best throughput. Others are lower but comparable amongst them.
- Throughput on VirtualBox is comparable to others when it is running a CPU intensive workload. Otherwise, VirtualBox throughput is significantly lower than others.
- Throughput of Xen, Native and LXC is similar in most workload cases.
- Throughput of LXC is significantly better than the rest when it is running a mixed workload type.

CPU and Memory Consumption. CPU and memory usage performance characteristics of various platforms are plotted in Fig. 5-6 for different workload types. Under constant load, mean memory usage on different platforms is shown in Fig. 5 for different workloads. Major observations are:

- LXC consumes significantly less memory than the rest for all workload scenarios.
- Native and Xen consume higher memory than others for all workload scenarios.
- VirtualBox consumes significantly less memory than Xen and Native for all workloads scenarios except DB-DB, where it consumes significantly higher than the rest.

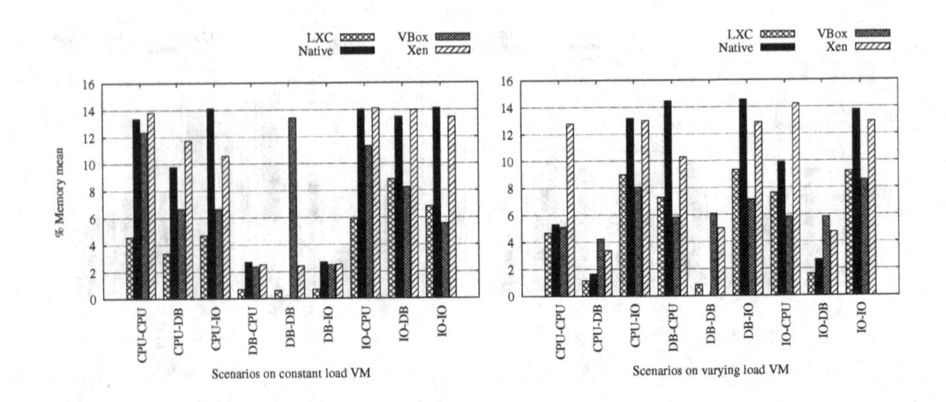

Fig. 5. Mean memory usage on VMs

Fig. 5 shows the mean memory consumption under increasing load. Key findings here are:

- Xen and Native consume higher memory in general for most cases.
- Xen consumes significantly higher memory than the rest for CPU-CPU scenario.
- LXC and VirtualBox consume least memory among all for all workload scenarios.

Mean CPU consumption on different platforms for different workload scenarios is depicted in Fig. 6 for a constant load. Following are the important observations from Fig. 6:

- For CPU intensive loads, VirtualBox used way higher CPU than the rest, for each of which the CPU usage was very small.
- For mixed load type, LXC consumed least CPU. Native and Xen were comparable but consumed more than LXC.
- For IO intensive workloads VirtualBox consumed least CPU. Others were comparable but consumed more than VirtualBox.

For an increasing load, the mean CPU consumption on different platforms is shown in Fig. 6 for various workload types. Key findings from Fig. 6 are as below:

- For IO intensive workloads, VirtualBox consumed the least CPU. Native consumed the most CPU for this case.
- For CPU intensive workloads, all platforms seem to consume very less and comparable amount of CPU.
- For mixed workloads, LXC consumes slightly more CPU than the rest.

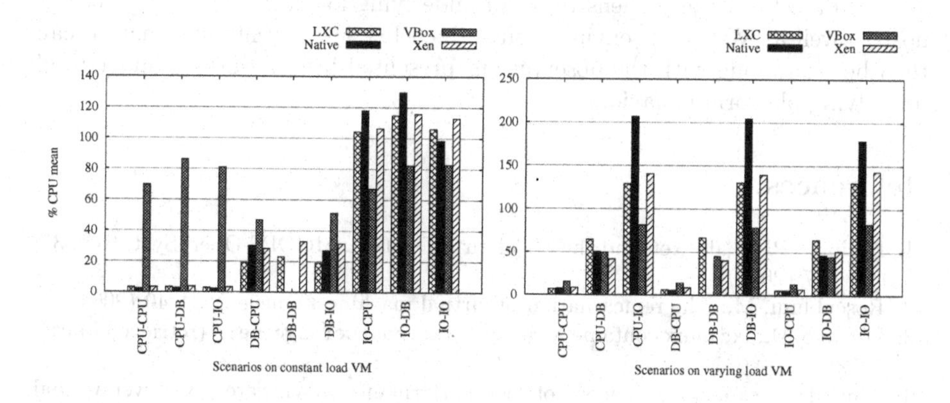

Fig. 6. Mean CPU usage on VMs (2 CPUs per VM)

Causes for Observed Behaviour. Linux perf tools [14] were used to watch some low level events on the host environment. Events such as instructions, context-switches, cache-misses, cache-references and page-faults reported by `perf stat` were observed. It was observed that for certain workload scenarios, virtualized platforms incurred greater number of cache misses per instruction, and it tended to increase with the load. This behaviour explains the cause for lower throughput in certain workload scenarios as indicated in previous sections. Nevertheless, there are several additional low level parameters and events to watch for in order to more precisely pin point the causes for observed performance characteristics; we intend to investigate those in future.

5 Conclusion and Future Work

The presented benchmarking results provide insights about the performance characteristics of various platforms for different workload scenarios. It allows finding suitable platforms for a given workload scenario. Presented results show that the impact of VM co-location on different performance indicators is dependent on the type of platform and the workload type. For instance, throughput experiences VM co-location impact for fewer platform-workload combinations. However, the impact of VM co-location on CPU and memory usage is visible for more number of scenarios. It was observed that performance of bare metal VMM and OS based virtualized platforms is comparable to the native platform in most scenarios. Hosted VMM based platform performance is comparable and even better to the others in a subset of scenarios, e.g., for IO intensive increasing loads, it consumes lesser CPU than others. We believe that the presented results prove useful for understanding the performance characteristics of different platforms under different workload scenarios.

A natural direction for future work is to examine and analyse the underlying causes for the observed performance characteristics. We would like to collect and analyse data for a comprehensive set of underlying low level events for physical host as well as VM guest environments. Such low level events information can then be seen along with the observations presented in this work to understand underlying platform behaviour.

References

1. Barham, P., et al.: Xen and the art of virtualization. SIGOPS Oper. Syst. Rev. 37, 164–177 (2003)
2. Rosenblum, M.: The reincarnation of virtual machines. Queue 2, 34–40 (2004)
3. Lezcano, D.: lxc linux containers, http://lxc.sourceforge.net/ (retrieved March 2012)
4. Ahmad, I., et al.: An analysis of disk performance in vmware esx server virtual machines. In: 2003 IEEE International Workshop on Workload Characterization, WWC-6 (2003)
5. Menon, A., et al.: Diagnosing performance overheads in the xen virtual machine environment. In: Proceedings of the 1st ACM/USENIX International Conference on Virtual Execution Environments (2005)
6. Padala, P., et al.: Performance evaluation of virtualization technologies for server consolidation. HP Laboratories Technical Report (2007)
7. Kolyshkin, K.: Virtualization in linux (openvz white paper) (2006), http://download.openvz.org/doc/openvz-intro.pdf
8. RUBiS Team, Rubis: Rice university bidding system. RUBiS Project, http://rubis.ow2.org/ (retrieved March 2012)
9. Watson, J.: Virtualbox: bits and bytes masquerading as machines. Linux Journal 2008 (February 2008), http://dl.acm.org/citation.cfm?id=1344209.1344210
10. Tomcat Project, Apache tomcat. The Apache Software Foundation, http://tomcat.apache.org/ (retrieved March 2012)
11. Hillyer, M.: Sakila sample database version 0.8. MySQL AB, http://dev.mysql.com/doc/sakila/en/sakila.html (retrieved March 2012)
12. JMeter Project, Apache jmeter. The Apache Software Foundation, http://jmeter.apache.org/ (retrieved March 2012)
13. GNU Octave, Statistics analysis, GNU, http://www.gnu.org/software/octave/doc/interpreter/Statistics.html (retrieved March 2012)
14. kernel.org, Linux profiling with performance counters. The Linux Kernel Archives, https://perf.wiki.kernel.org/ (retrieved March 2012)

Open Execution Engines of Stream Analysis Operations

Qiming Chen and Meichun Hsu

HP Labs
Palo Alto, California, USA
Hewlett Packard Co.
{qiming.chen,meichun.hsu}@hp.com

Abstract. In this paper we describe our massively *parallel* and *elastic* stream analysis platform; it is capable of supporting the graph-structured dataflow process with each logical operator executed by multiple physical instances running in parallel over distributed server nodes. We propose the **canonical dataflow operator** framework to provide automated and systematic support for executing, parallelizing and granulizing the continuous operations.

We focus on the following issues: first, how to categorize the meta-properties of stream operators such as the I/O, blocking, data grouping characteristics, for providing unified and automated system support; next, how to elastically and correctly parallelize a stateful operator that is history-sensitive, relying on the prior state and data processing results; and further, how to analyze unbounded stream granularly to ensure sound semantics (e.g. aggregation). These issues are not properly abstracted and systematically addressed in the current generation of stream processing systems, but left to user programs which can result in fragile code, disappointing performance and incorrect results.

We tackle these issues by introducing the ***open-executors***. An open executor supports the streaming operations with specific characteristics and running pattern, but is *open* for the application logic to be plugged-in. We illustrate the power of this approach by showing the system support in parallelizing and granulizing dataflow operations *safely* and *correctly*. The proposed canonical operation framework can be generalized to allow us to standardize various operational patterns of stream operators, and have these patterns supported systematically and automatically. We have built this platform; our experience reveals its value in real-time, continuous, elastic data-parallel and topological stream analysis process.

1 Introduction

Real-time stream analytics has increasingly gained popularity since enterprises need to capture and update business information just-in-time, analyze continuously generated "moving data" from sensors, mobile devices, social media of all types, and gain live business intelligence.

We have built a stream analytics platform with code name ***Fontainebleau*** for dealing with *continuous, real-time* data-flow with *graph-structured* topology. This platform is massively *parallel, distributed* and *elastic* with each logical operator executed by multiple physical instances running in parallel over distributed server nodes. The stream

A. Hameurlain et al. (Eds.): Globe 2012, LNCS 7450, pp. 75–87, 2012.

analysis operators are defined by users flexibly. From stream abstraction point of view, our stream analytics cluster is positioned in the same space of System S(IBM), Dryad(MS), Storm(Tweeter), etc. However, this work aims to advance the state of art by providing canonical execution support for stream analysis operators.

1.1 The Challenges

A stream analytics process with continuous, graph-structured dataflow is composed by multiple operators and the pipes connecting these operators. The operators for stream analysis have certain meta-properties representing their I/O characteristics, blocking characteristics, data grouping characteristics, etc, which can be categorized for introducing unified system support. Categorizing stream operators and their running patterns to provide automatic support accordingly, can ensure the operators to be executed optimally and consistently, as well as ease user's effort for dealing with these properties manually which is often tedious and risky. Unfortunately, this issue has been missed by the existing stream processing systems.

Next, to scale out, the data-parallel execution of operators must be taken into account, where how to ensure the correctness of data-parallelism is the key issue, and requires the appropriate system protocol to guarantee; particularly in parallelizing stateful stream operators where the stream data partitioning and data buffering must be consistent.

Further, stream processing is often made in granule. For example, to provide sound aggregation semantics (e.g. sum), the infinite input data stream must be processed chunk by chunk where each operator may punctuate data based on different chunking criteria such as in 1-minute or 1-hour time windows (certain constraints apply, e.g. the frame of a downstream operator must be the same as, or some integral number of, the frame of its upstream operator). Granulizing dataflow analytics represents another kind of common behavior of stream operators which also need to be supported systematically.

Current large-scale data processing tools, such as Map-Reduce, Dryad, Storm, etc, do not address these issues in a canonical way. As a result, the programmers have to deal with them on their own, which can lead to fragile code, disappointing performance and incorrect results.

1.2 The Proposed Solution

The operators on a parallel and distributed dataflow infrastructure are performed by both the infrastructure and the user programs, which we refer to as their **template behavior** and **dynamic behavior**. The template behavior of a stream operator depends on its meta-properties and its running pattern. For example, a map-reduce application is performed by the Hadoop infrastructure as well as the user-coded map function and reduce function. Our streaming platform is more flexible and elastic than Hadoop in handling dynamically parallelized operations in a general graph structured dataflow topology, and our focus is placed on supporting the template behavior, or operation patterns, **automatically** and **systematically**.

Unlike applying an operator to data, stream processing is characterized by the flowing of data through a *stationed* operator. We introduce the notion of **open-station** as the container of a stream operator. The stream operators with certain common meta-properties can be executed by the class of open-stations specific to these operators. Open-stations are classified into a station hierarchy. Each class provides an **open-executor** as well as related system utilities. In the OO programming context, the open-executor is coded by invoking certain abstract functions (methods) to be implemented by users based on their application logic. In this way the station provides designated system support, while *open* for the application logic to be plugged-in. In this work we use the proposed architecture to solve several typical stream processing problems.

The key to ensure safe parallelization is to handle data flow group-wise - for each vertex representing a logical operator in the dataflow graph; the operation parallelization with multiple instances comes with input data partition (grouping) which is consistent with the data buffering at each operation instance. This ensures that in the presence of multiple execution instances of an operator, O, every stream tuple is processed *once and only once* by one of the execution instances of O; the historical data processing states of every group of the partitioned data are buffered with *one and only one execution instance* of O. Our solution to this problem is based on the open station architecture.

The key to ensure the granule semantics is to handle dataflow chunk wise by punctuating and buffering data consistently. Our solution to this problem is also based on the open station architecture.

In general, the proposed canonical operation framework allows us to *standardize* various operational patterns of stream operators, and have these patterns supported systematically and automatically. Our experience shows its power in real-time, continuous, elastic data-parallel and topological stream analytics.

The rest of this paper is organized as follows: section 2 describes the notions of open-station and open-executor; then based on these notions section 3 discusses how to guarantee the correctness of data-parallel execution of stateful operations, and section 4 deals with the granular execution of stream operations. The experimental results are illustrated in section 5. Finally section 6 compares with related work and concludes the paper.

2 Open Station and Open Executor of Stream Operator

2.1 Continuous, Parallel and Elastic Stream Analytics Platform

Fontainebleau is a real-time, continuous, parallel and elastic stream analytics platform. There are two kinds of nodes on the cluster: the *coordinator node* and the *agent nodes* with each running a corresponding daemon. A dataflow process is handled by the coordinator and the agents spread across multiple machine nodes. The coordinator is responsible for distributing code around the cluster, assigning tasks to machines, and monitoring for failures, in the way similar to Hadoop's job-tracker. Each agent interacts with the coordinator and executes some operator instances (as threads) of the dataflow process. The *Fontainebleau* platform is built using several open-source tools, including ZooKeeper[12], ØMQ[11], Kryo[13], Storm[14], etc. ZooKeeper coordinates distributed applications on multiple nodes elastically. ØMQ support efficient and reliable messaging. Kryo deals with object serialization. Storm provides the basic dataflow topology support.

A stream is an unbounded sequence of tuples. A stream operator transforms a stream into a new stream based on its application-specific logic. The graph-structured stream transformations are packaged into a "topology" which is the top-level dataflow process. When an operator emits a tuple to a stream, it sends the tuple to every successor operators subscribing to that stream. A stream grouping specifies how to group and partition the tuples input to an operator. There exist a few different kinds of stream groupings such as hash-partition, replication, random-partition, etc.

To support elastic parallelism, we allow a logical operator to execute by multiple physical instances, as threads, in parallel across the cluster, and they pass messages to each other in a distributed way. Using the ØMQ library [11], message delivery is reliable; messages never pass through any sort of central router, and there are no intermediate queues.

Compared to Hadoop, first, our platform is characterized by "**real-time**" and "**continuous**", with the capability of parallel and distributed computation on real-time and infinite streams of messages, events and signals. Next, it is characterized by "**topological**" in the sense that it deals with dataflow in the complex graph-structured topology, not limited to the map-reduce scheme. Finally, unlike a statically configured Hadoop platform, the Fontainebleau platform can scaled-out over a grid of computers "**elastically**" for parallel computing.

We use a simplified as well as extended Linear-Road (LR) benchmark to illustrate the notion of stream process. The LR benchmark models the traffic on 10 express ways; each express way has two directions and 100 segments. Cars may enter and exit any segment. The position of each car is read every 30 seconds and each reading constitutes an event, or stream element, for the system. A car position report has attributes *vehicle_id*, *time* (in seconds), *speed* (mph), *xway* (express way), *dir* (direction), *seg* (segment), etc. With the simplified benchmark, the traffic statistics for each highway segment, i.e. the number of active cars, their average speed per minute, and the past 5-minute moving average of vehicle speed, are computed. Based on these per-minute per-segment statistics, the application computes the tolls to be charged to a vehicle entering a segment any time during the next minute. As an extension to the LR application, the traffic statuses analyzed and reported every hour. The logical stream process for this example is given in Fig. 1.

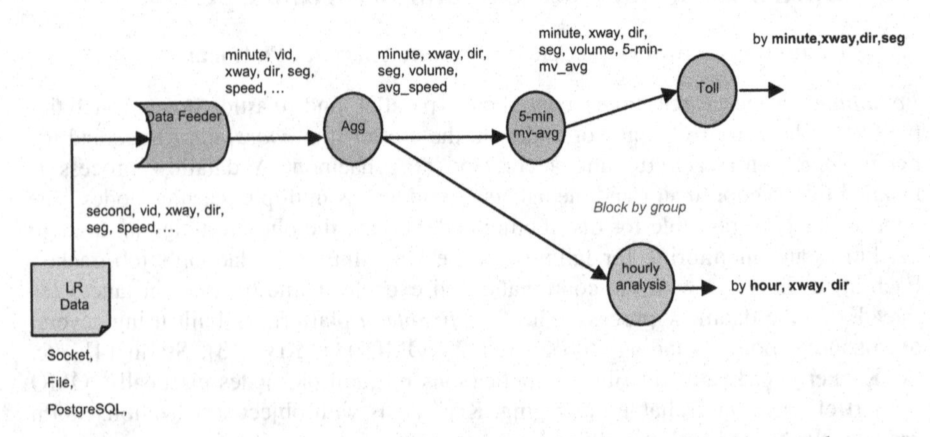

Fig. 1. The extended LR logical dataflow process with operators linked in a topology

This stream analytics process is specified by the Java program illustrated below.

```
public class LR_Process {
...
public static void main(String[] args) throws Exception {
    ProcessBuilder builder = new ProcessBuilder();
    builder.setFeederStation("feeder", new LR_Feeder(args[0]), 1);
    builder.setStation("agg", new LR_AggStation(0, 1), 6) .hashPartition("feeder",
      new Fields("xway", "dir", "seg"));
    builder.setStation("mv", new LR_MvWindowStation(5), 4).hashPartition("agg",
      new Fields("xway", "dir", "seg"));
    builder.setStation("toll", new LR_TollStation(), 4).hashPartition("mv",
      new Fields("xway", "dir", "seg"));
    builder.setStation("hourly", new LR_BlockStation(0, 7), 2).hashPartition("agg",
      new Fields("xway", "dir"));
    Process process = builder.createProcess();
    Config conf = new Config();  conf.setXXX(...); ...
    Cluster cluster = new Cluster();
    cluster.launchProcess("linear-road", conf, process);
    ...
  }
```

In the above topology specification, the hints for parallelization are given to the operators "agg" (6 instances), "mv" (5 instances), "toll" (4 instances) and "hourly" (2 instances), the platform may make adjustment based on the resource availability. Then the physical instances of these operators for data-parallel execution are illustrated in Fig 2.

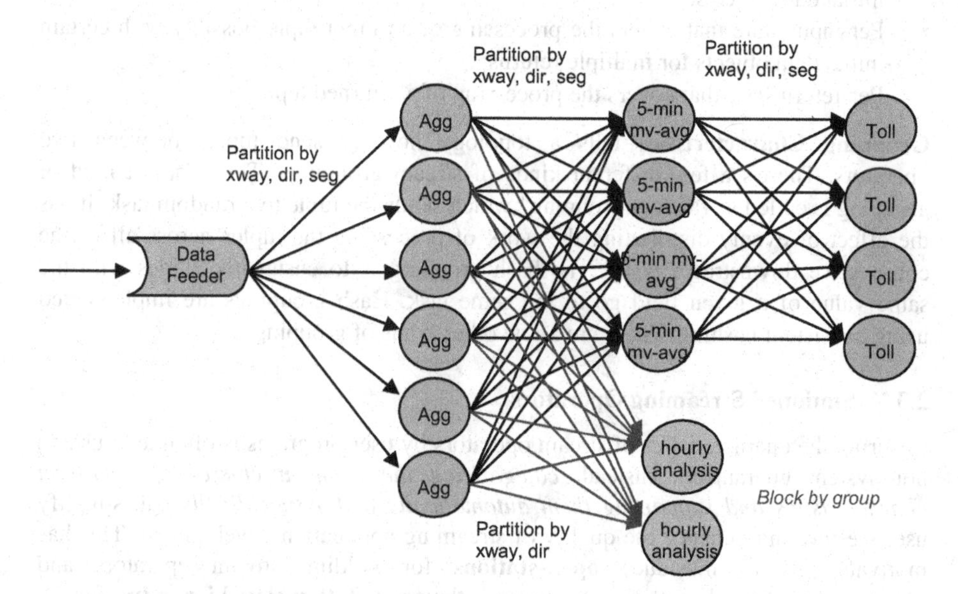

Fig. 2. The LR dataflow process instance with elastically parallelized operator instances

2.2 Meta Characteristics of Operators

Stream operators have certain characteristics in several dimensions, such as the provisioning of initial data, the granularity of event processing, memory context, invocation patterns, results grouping and shuffling, etc, which may be considered as the meta-data, or the design pattern of operators. Below we briefly list some characteristics.

I/O Characteristics specifies the number of input tuples and the output tuples the stream operator is designed to handle the stream data *chunk-wise*. Examples are 1:1 (one input/one output), 1:N (one input/multiple outputs), M:1(multiple inputs/ one output) and M:N (multiple inputs/ multiple outputs). Accordingly we can classify the operators into Scalar (1:1); Table Valued (TV) (1:N); Aggregate (N:1), etc, for each chunk of the input. Currently we support the following chunking criteria for punctuating the input tuples: (a) by cardinality, i.e. number of tuples; and (b) by granule as a function applied to an attribute value, e.g. *get_minute* (timestamp in second).

Blocking Characteristics tells that in the multiple input case, the operator applies to the input tuple one by one incrementally (e.g. per-chunk aggregation), or first pools the input tuples and then apply the function to all the pooled tuples. Accordingly the block mode can be *per-tuple* or *blocking*. Specifying the blocking characteristics tells the system to invoke the operator in the designated way, and save the user's effort to handle them in the application program.

Caching Characteristics is related to the 4 levels potential cache states:

- per-process state that covers the whole dataflow process with certain initial data objects;
- Per-chunk state that covers the processing of a chunk of input tuples with certain initial data objects;
- Per-input state that covers the processing of an input tuple possibly with certain initial data objects for multiple returns;
- Per-return state that covers the processing of a returned tuple.

Grouping Characteristics tells a topology how to send tuples between two operators. There's a few different kinds of stream groupings. The simplest kind of grouping is called a "random grouping" which sends the tuple to a random task. It has the effect of evenly distributing the work of processing the tuples across all of the consecutive downstream tasks. The hash grouping is to ensure the tuples with the same value of a given field go to the same task. Hash groupings are implemented using consistent hashing. There are a few other kinds of groupings.

2.3 Stationed Streaming Operators

Ensuring the characteristics of stream operators by user programs is often tedious and not system guaranteed. Instead, *categorizing the common classes of operation characteristics and supporting them automatically and systematically* can simplify user's effort and enhance the quality of streaming application development. This has motivated us to introduce **open-stations** for holding stream operators and encapsulating their characteristics – towards the **open station class hierarchy** (Fig 3).

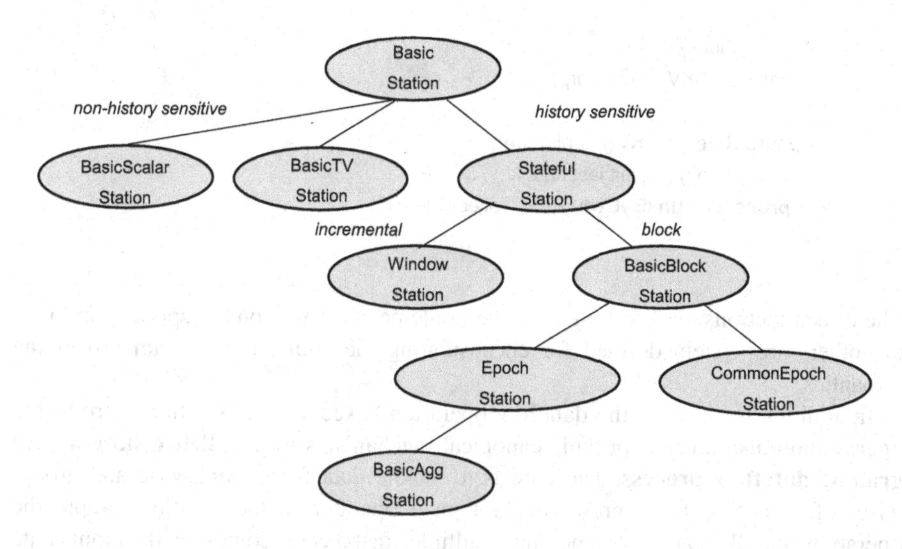

Fig. 3. Station Hierarchy Example

Each open-station class is provided with an "open executor" as well as related system utilities for executing the corresponding kind of operators; that "open" executor invokes the abstract methods, which are defined in the station class but implemented by users with the application logic. In this way a station provides the designated system support, while open for the application logic to be plugged-in. A user defined operator captures the designated characteristics by subclass the appropriate station, and captures the application logic by implementing the abstract methods accordingly.

2.4 Open Executor

Specified in a station class, there are two kinds of pre-prepared methods: the system defined ones and the user defined ones.

- The system defined methods include the **open-executor** and other utilities which is open to plugging-in application logic, in the sense that they invoke the abstract methods to be implemented by users according to the application logic.
- The abstract methods to be implemented by the user based on the application logic.

For example, the EpochStation that extends BasicBlockStation, is used to support chunk-wise stream processing, where the framework provided functions, hidden from user programs, include

```
public boolean nextChunk(Tuple, tuple) {// group specific ...}
public void execute(Tuple tuple, BasicOutputCollector collector) {
    boolean new_chunk = nextChunk(tuple);
    String grp = getGroupKey(tuple);
    GroupMeasures gm = null;
```

```
        if (new_chunk) {
            gm = getGKV().dump(grp);
        }
        updateState(getGKV(), tuple, grp);
        if (new_chunk) { //emit last chunk
            processChunkByGroup(gm, collector);
        }
    }
```

The three functions marked bolt are to be implemented based on the application logic; the others are system defined for encapsulating the chunk-wise stream processing semantics.

In addition to offering the dataflow operation "executor" abstraction, introducing open station also aims to provide canonical mechanisms to **parallelize stateful and granule dataflow process**. The core is to handle data flow chunk-wise and group-wise - for each vertex representing a logical operator in the dataflow graph; the operation parallelization (launching multiple instances) comes with input data partition (grouping) which is consistent with the data buffering at each operation instance. These are discussed in the following sections.

3 Group-Wise Data-Parallel Execution of Stateful Tasks

3.1 Elastic Data-Parallel Execution of Operators

Under our approach, logically, the dataflow elements, i.e. tuples, either originated from a data-source or derived by a logical operator, say *A*, are sent to one or more receiving logical operators, say *B*. However, each logical operator may have multiple execution instances, the dataflows from *A* to *B* actually form multi-to-multi messaging.

To handle data-parallel operations, an operator property: parallelism hint, can be specified, that is the number (default to 1) of station threads for running the operator. The number of actual threads will be judged by the infrastructure and load-balanced over multiple available machines.

For the sake of correct parallelism, the stream from *A*'s instances to *B*'s instances are sent in a *partitioned* way (e.g. hash-partition) such that the data sent from any instance of *A* to the instances of *B* are partitioned in the same way. This is similar to the data shuffling from a Map node to a Reduce node, but in more general dataflow topology.

Although our platform offers the flexibility of dataflow grouping with options hash-partition, random-partition, range-partition, replicate, etc, the platform enforces the use of hash partition for the parallelized operators. In case an operator is specified to have parallel instances in the user's dataflow process specification, the input stream to that operator must be defined as hash-partitioned; otherwise the process specification would be invalidated.

Further, there can be multiple logical operators, B_1, B_2, …, B_n, for receiving the output stream of *A*, but each with different data partition criterion, called *inflow-*

grouping-attributes (a la SQL group by). The tuples falling in the same partition, i.e. grouped together, have the same ***"inflow-group-keys"***. For example, the tuples representing the traffic status of an express way (*xway*), direction (*dir*) and segment (*seg*), are partitioned, thus grouped by attributes <xway, dir, seg>; tuples of each group has the same inflow-group-key derived from the values of xway, dir and seg. An operation instance may receive multiple groups of data. The abstract method, ***getGroupKey***(tuple), must be implemented, which is invoked by the corresponding open-executor.

3.2 Parallelize Stateful Streaming Operators Group-Wise

A stateful operator caches its state for future computation, and therefore is history sensitive. When a logical stateful operator has multiple instances, their input data must be partitioned, and the data partition must be consistent with the data buffering.

For example, given the logical operation, *O*, for calculating moving-average and with the input stream data partitioned by <xway, dir, seg>, the data buffers of its execution instances are also partitioned by <xway, dir, seg>, which is prepared and enforced by the system.

For history-sensitive data-parallel computation, an operation instance keeps a state computed from its input tuples (other static states may be incorporated but not the focus of this discussion). We generally provide this state as a KV store where keys, referred to as *cachhing-group-keys*, are Objects (e.g. String) extracted from the input tuples, and values are Objects derived from the past and present tuples such as numerical objects (e.g. sum, count), list objects (certain values derived from each tuple), etc. the multiple instances of a logical operation can run in data-parallel provided that the inflow-group-keys are used as the caching group-keys. In this sense we refer to the KV store as Group-wise KV store (GKV). APIs for accessing the GKV are provided as well. As illustrated in the last section, an important abstract method, `updateState()`, is defined and to be implemented by users.

With the above mechanisms, in the presence of multiple execution instances of an operator, every stream tuple is processed **once and only once** by one of the execution instances; the data processing states of every group of the partitioned input data (e.g. the tuples belonging to the same segment of the an express-way in a direction) are buffered in the function closure of **one and only one execution instance** of that operator. These properties are common to a class of tasks thus we support them in the corresponding station class, that, substantially, is subclassifiable.

4 Epoch Based Stream Analytics

Although a data stream is unbounded, very often applications require those infinite data to be analyzed granularly. Particularly, when the stream operation involves the aggregation of multiple events, for semantic reason the input data must be punctuated into bounded chunks. This has motivated us to execute such operation *epoch by epoch* to process the stream data *chunk by chunk*.

For example, in the previous car traffic example, the operation "agg" aims to deliver the average speed in each express-way's segment per minute. Then the execution of this operation on an infinite stream is made in a sequence of *epochs*, one on each stream chunks. To allow this operation to apply to the stream data one chunk at a time, and to return a sequence of chunk-wise aggregation results, the input stream, is cut into 1 minute (60 seconds) based chunks, say S_0, S_1, ...S_i, ... such that the execution semantics of "agg" is defined as a sequence of one-time aggregate operation on the data stream input minute by minute.

In general, given an operator, O, over an infinite stream of relation tuples S with a criterion ϑ for cutting S into an unbounded sequence of chunks, e.g. by every 1-minute time window, $<S_0, S_1, \ldots, S_i, \ldots>$ where S_i denotes the *i-th* "chunk" of the stream according to the chunking-criterion ϑ. The semantics of applying O to the unbounded stream S lies in

$$Q(S) \rightarrow < Q(S_0), \ldots Q(S_i), \ldots >$$

which continuously generates an unbounded sequence of results, one on each *chunk* of the stream data.

Punctuating input stream into chunks and applying operation *epoch by epoch* to process the stream data *chunk by chunk*, is a template behavior common to many stream operations, thus we consider it as a kind of meta-property of a class of stream operations and support it automatically and systematically by our operation framework. In general, we host such operations on the **epoch station** (or the ones subclassing it) and provide system support in the following aspects (please refer to the epoch station example given previously).

- An epoch station hosts a stateful operation that is data-parallelizable, and therefore the input stream must be hash-partitioned which is consistent with the buffering of data chunks as described in the last section.
- Several types of stream punctuation criteria are specifiable, including punctuation by cardinality, by time-stamps and by system-time period, which are covered by the system function

  ```
  public boolean nextChunk(Tuple, tuple)
  ```

  to determine whether the current tuple belongs to the next chunk or not.
- If the current tuple belongs to the new chunk, the present data chunk is dumped from the chunk buffer for aggregation/group-by in terms of the user-implemented abstract method `processChunkByGroup()`.
- Every input tuple (or derivation) is buffered, either into the present or the new chunk.

By specifying additional meta properties and by subclassing the epoch station, more concrete system support can be introduced. For example, an aggregate of a chunk of stream data can be made once by end of the chunk, or tupe-wise incrementally. In the latter case an abstract method for per-tuple updating the partial aggregate is provided and implemented by the user.

The *paces of dataflow* wrt timestamps can be different at different operators; for instance, the "agg" operator is applied to the input data minute by minute, so are some downstream operators of it; however the "hourly analysis" operator is applied to the input stream minute by minute, but generates output stream elements hour by hour.

The combination of group-wise and chunk-wise stream analytics provides a generalized abstraction for parallelizing and granulizing the continuous and incremental dataflow analytics.

5 Experiments

We have built the Fontainebleau prototype based on architecture and policies explained in the previous sections. In this section we briefly overview our experimental results. Our testing environment include 16 Linux servers with gcc version 4.1.2 20080704 (Red Hat 4.1.2-50), 32G RAM, 400G disk and 8 Quad-Core AMD Opteron Processor 2354 (2200.082 MHz, 512 KB cache). One server holds the coordinator daemon, 15 other servers hold the agent daemons, each agent supervises several worker processes, and each worker process handles one or more task instances. Based on the topology and the parallelism hint of each logical task, one or more instances of that task will be instantiated by the framework to process the data streams.

We first present the experiment results of running the above example topology that is similar to the Linear Road scenario; our topology extends that scenario but we use the same test data under the stress test mode - the data are read from a file continuously without following the real-time intervals, leading to a fairly high throughput. Below are our findings:

The performance show in Fig. 4 is based on the event rate of 1.33 million per minute with approximate 12 million (11,928,635) input events. Most of the tasks have 28 parallel instances except one having 14 parallel instances. There is no load-shedding (dropping events) observed.

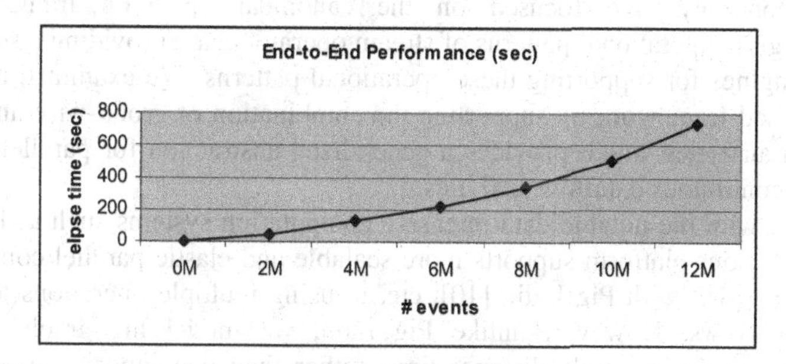

Fig. 4. The performance of data-parallel stream analytics with the LR topology

Besides performance, we are particularly interested in the load shedding behavior of the system. Load shedding is the normal behavior of stream processing with different policies studied, such as event sampling. However, for certain kind of applications, event dropping is not tolerable, and then the parallelization of tasks comes to the picture. Our experience shows, given an event generation task S and the successor task T, to avoid event drop, the multiplication of T's throughput and the number of parallel instances of T, should be higher than S's throughput. This also applies to each pair of predecessor and successor tasks.

With our test, in case no task parallelism supported (every task has just 1 instance), only 11,681,204 events are processed, meaning that the load shedding rate is approximate 2%. Such a load shedding rate is limited due to the computation for event formatting is rather heavy comparable with the subsequent event processing tasks. We observe that if the throughput of event generation task is high, and/or if the first task after it (or any task that needs to process every message from it) cannot match the throughput of it, then message dropping is observable.

Fig 5 shows that when parallelism increases, the effect of load shedding, or dropping messages, can be reduced or possibly eliminated. On this issue we are investigating further architecture solutions.

	# nodes	# agg instances	# mv instances	# toll instances	# hourly instances	# original events	# processed events	event loss- rate
A	6	1	1	1	1	11,928,635	11,681,204	2%
B	14	14	14	14	4	11,928,635	11,809,348	1%
C	14	28	28	28	14	11,928,635	11,928,635	0%
D	14	56	56	56	14	11,928,635	11,928,635	0%

Fig. 5. Parallelism vs. load shedding

6 Related Work and Conclusions

In this paper we described our parallel and distributed stream analysis system capable of executing the real-time, continuous streaming process with general graph-structured topology. We focused on the canonical operation framework for standardizing the operational patterns of stream operators, and providing a set of open execution engines for supporting these operational patterns. We examined the power of the proposed framework by supporting the combination of group-wise and chunk-wise stream analytics which provides a generalized abstraction for parallelizing and granulizing continuous dataflow analytics.

Compared with the notable data-intensive computation systems such as DISC [3], Dryad [8], etc, our platform supports more scalable and elastic parallel computation. We share the spirit with Pig Latin [10], etc, in using multiple operations to express complex dataflows. However, unlike Pig Lain, we model the graph structured dataflow by composing multiple operations rather than decomposing a query into multiple operations; our data sources are dynamic data streams rather than static files; we partitioning stream data on the fly dynamically, rather than prepare partitioned files statically to Map-Reduce them. This work also extends the underlying tools such as Storm by elaborating it from a computation infrastructure to a state conscious computation/caching infrastructure, and from the user task oriented system to the execution engine oriented system.

Supporting truly continuous operations distinguish our platform from the current generation of stream processing systems, such as System S (IBM), STREAM (Stanford) [1], Aurora, Borealis[2], TruSQL[9], etc.

Compared with Hadoop, the Fontainebleau platform has the following advantages.

- Fontainebleau deals with arbitrary graph-structured dataflow topology
- Fontainebleau deals with group-wise operation for any task, and multiple logical operations can specify different inflow-grouping-attributes from the output stream of an upstream logical operation. In this way we generalized the data partition spirit of MR.
- Fontainebleau deals with chunk-wise computation with different pace at different operations.
- Hadoop supports data-parallelism by the fixed infrastructure, while we do it by the canonical operator framework. Our mechanisms raise a programming abstraction that makes it easy to express incremental computations over incrementally-arriving data; and these mechanisms are geared specifically toward continuous incremental workloads.

Envisaging the importance of standardizing the operational patterns of dataflow operators, we are providing a rich set of open execution engines and linking stations with existing data processors such as DBMS and Hadoop, towards the integrated dataflow cloud service.

References

[1] Arasu, A., Babu, S., Widom, J.: The CQL Continuous Query Language: Semantic Foundations and Query Execution. VLDB Journal 15(2) (June 2006)

[2] Abadi, D.J., et al.: The Design of the Borealis Stream Processing Engine. In: CIDR (2005)

[3] Bryant, R.E.: Data-Intensive Supercomputing: The case for DISC. CMU-CS-07-128 (2007)

[4] Chen, Q., Hsu, M., Zeller, H.: Experience in Continuous analytics as a Service (CaaaS). In: EDBT 2011 (2011)

[5] Chen, Q., Hsu, M.: Experience in Extending Query Engine for Continuous Analytics. In: Bach Pedersen, T., Mohania, M.K., Tjoa, A.M. (eds.) DAWAK 2010. LNCS, vol. 6263, pp. 190–202. Springer, Heidelberg (2010)

[6] Chen, Q., Hsu, M.: Continuous MapReduce for In-DB Stream Analytics. In: Meersman, R., Dillon, T., Herrero, P. (eds.) OTM 2010. LNCS, vol. 6428, pp. 16–34. Springer, Heidelberg (2010)

[7] Dean, J.: Experiences with MapReduce, an abstraction for large-scale computation. In: Int. Conf. on Parallel Architecture and Compilation Techniques. ACM (2006)

[8] Isard, M., Budiu, M., Yu, Y., Birrell, A., Fetterly, D.: Dryad: Distributed data-parallel programs from sequential building blocks. In: EuroSys 2007 (March 2007)

[9] Franklin, M.J., et al.: Continuous Analytics: Rethinking Query Processing in a Network¬Effect World. In: CIDR 2009 (2009)

[10] Olston, C., Reed, B., Srivastava, U., Kumar, R., Tomkins, A.: Pig Latin: A Not-So-Foreign Language for Data Processing. In: ACM SIGMOD 2008 (2008)

[11] ØMQ Lightweight Messaging Kernel, http://www.zeromq.org/

[12] Apache ZooKeeper, http://zookeeper.apache.org/

[13] Kryo - Fast, efficient Java serialization, http://code.google.com/p/kryo/

[14] Twitter's Open Source Storm Finally Hits, http://siliconangle.com/blog/2011/09/20/twitter-storm-finally-hits/

Improving Content Delivery by Exploiting the Utility of CDN Servers

George Pallis

Computer Science Department
University of Cyprus
gpallis@cs.ucy.ac.cy

Abstract. As more aspects of our work and life move online and the Internet expands beyond a communication medium to become a platform for business and society, Content Delivery Networks (CDNs) have recently gained momentum in the Internet computing landscape. Today, a large portion of Internet traffic is originating from CDNs. The ultimate success of CDNs requires novel policies that would address the increasing demand for content. Here we exploit the CDN utility - a metric that captures the traffic activity in a CDN, expressing the usefulness of surrogate servers in terms of data circulation in the network. We address the content replication problem by replicating content across a geographically distributed set of servers and redirect users to the closest server in terms of CDN utility. Through a detailed simulation environment, using real and synthetically generated data sets we show that the proposed approach can improve significantly the performance of Internet-based content delivery.

Keywords: Internet-based Content Delivery, Replication, CDN pricing.

1 Introduction

Content Delivery Networks (CDNs) have emerged to overcome the inherent limitations of the Internet in terms of user perceived Quality of Service (QoS) when accessing Web data. They offer infrastructure and mechanisms to deliver content and services in a scalable manner, and enhance users' Web experience. Specifically, a CDN is an overlay network across the Internet, which consists of a set of servers (distributed around the world), routers and network elements. Edge servers are the key elements in a CDN, acting as proxy caches that serve directly cached content to users. With CDNs, content is distributed to edge cache servers located close to users, resulting in fast, reliable applications and Web services for the users. Once a user requests content on a Web provider (managed by a CDN), the user's request is directed to the appropriate CDN server. The perceived high end-user performance and cost savings of using CDNs have already urged many Web entrepreneurs to make contracts with CDNs. For instance, Akamai - one of the largest CDN providers in the world - claims to be delivering 20% of the world's Web traffic. While the real numbers are debatable, it is clear that CDNs play a crucial role in the modern Internet infrastructure.

A. Hameurlain et al. (Eds.): Globe 2012, LNCS 7450, pp. 88–99, 2012.

In [10], we introduced a metric, called CDN utility metric, in order to measure the utility of CDN servers. A similar metric has also been used in [12] for a p2p system. Recent works [8,9,13] have shown that CDN utility metric captures the traffic activity in a CDN, expressing the usefulness of CDN servers in terms of data circulation in the network. In particular, the CDN utility is a metric that expresses the relation between the number of bytes of the served content against the number of bytes of the pulled content (from origin or other CDN servers). In this work we go one step further and propose a utility-based replication policy for CDNs. Maximizing the CDN utility metric, we can both reduce network utilization and overall traffic volume, which may further reduce network congestion and thus improve the overall user performance. Using an extensive simulation testbed, we show that this metric can improve the CDN's performance. Our contributions in this paper can be summarized as follows:

- We formulate the problem of optimally replicating the outsourced content in surrogate servers of a CDN.
- We present a novel adaptive technique under a CDN framework on which we replicate content across a geographically distributed set of servers and redirect users to the closest server in terms of CDN utility. Maximizing the CDN utility, we maximize the usefulness of CDN surrogate servers and minimize the final cost of Web content providers.
- We develop an analytic simulation environment to test the efficiency of the proposed scheme. Using real and synthetically generated data sets, we show the robustness and efficiency of the proposed method which can reap performance benefits.

In this work, we use the CDN utility metric which has been introduced in [10], in order to determine the placement of outsourced content to CDN surrogate servers. Considering that CDN utility expresses the relation between the number of bytes of the served content against the number of bytes of the pulled content (from origin or other surrogate servers), maximizing this value would improve content delivery.

The structure of the paper is as follows. In Section 2, we formulate the content replication problem for improving content delivery and in Section 3 the proposed approach is described. Sections 4 and 5 present the simulation testbed and the evaluation results respectively. Section 6 presents an overview of related work and Section 7 concludes the paper.

2 Problem Formulation

In CDNs, the content is pushed (proactively) from the origin Web server to CDN surrogate servers and then, the surrogate servers cooperate in order to reduce the replication and update cost.

We consider a popular Web site that signs a contract with a CDN provider with N surrogate servers, each of which acts as an intermediary between the servers and the end-users. We further assume that the surrogate server i has S_i bytes of storage capacity, where $i \in \{1, ..., N\}$.

In order to formulate the placements cost function, we assume that we have K requested objects. Each object k has a size of s_k, where $k \in \{1, ..., K\}$. In this context, we define a variable which determines if an object k is stored to surrogate server i as follows:

$$f_{ik} = 1 \quad (1)$$

if object k is stored at surrogate i, else $f_{ik} = 0$.

The storage is subject to the constraint that the space available at surrogate server i is bounded by $\sum_{k=1}^{K} s_k f_{ik} \leq S_i$, where $i \in \{1, ..., N\}$. Furthermore, each surrogate server can hold at most one replica of the object.

Considering that all the requested objects are initially placed on an origin server (the initial placement is denoted by x_o), the content replication problem is to select the optimal placement x (defines the placement of requested objects to CDN surrogate servers) such that it minimizes:

$$cost(x) = \sum_{i=1}^{N} \sum_{k=1}^{K} p_{ik}(D_{ik}(x)) \quad (2)$$

where $D_{ik}(x)$ is the "distance" to a replica of object k from surrogate server i under the placement x and p_{ik} is the likelihood that the object k will be requested by server i. This likelihood encapsulates the users' satisfaction.

However, as it has been proved in previous studies, this problem is NP complete (it is similar to the NP-complete knapsack problem), which means that for a large number of requested objects and surrogate servers is not feasible to solve this problem optimally. In this context, we propose a new heuristic strategy where its criterion is the CDN utility metric.

3 The CDN Utility Replica Placement Algorithm

The main idea is to place the requested objects to surrogate servers with respect to CDN utility metric. The intuition of this metric is that a surrogate server is considered to be useful (high net utility) if it uploads content more than it downloads, and vice versa. It is bounded to the range [0..1] and provides an indication about the CDN traffic activity.

Formally, the CDN utility u is expressed as the mean of the individual utilities of each surrogate server [10]. Considering that a CDN has N surrogate servers, the CDN utility u can be defined as follows:

$$u = \frac{\sum_{i=1}^{N} u_i}{N} \quad (3)$$

where u_i is the net utility of a CDN surrogate server i and it is quantified by using the following equation:

$$u_i = \frac{2}{\pi} \times arctan(\xi) \quad (4)$$

Algorithm 1. The CDN Utility Replica Placement Algorithm

Input: $O[1..K]$: users' requested objects;
Input: $S[1..N]$: CDN surrogate servers;
Output: a placement x of requested objects to surrogate servers

```
1: for all obj ∈ O do
2:    for all s ∈ S do
3:       compute the CDN-utility[obj][s];
4:    end for
5: end for
6: for all obj ∈ O do
7:    for all s ∈ S do
8:       create a list L: each element of the list includes the pairs of outsoursed object
         surrogate server ordering with respect to the maximum CDN-utility;
9:    end for
10: end for
11: while (list L not empty AND cache size of S not full) do
12:    get the first element from L: e=(s-obj, s-s);
13:    if (size of s-obj ≤ cache size of s-s) then
14:       place s-obj to s-s;
15:    end if
16:    delete e from L;
17: end while
```

Fig. 1. The CDN Utility Replica Placement Algorithm

The parameter ξ is the ratio of the uploaded bytes to the downloaded bytes. The arctan function in (4) assists to obtain scaled resulting utility in the range [0..1]. The value $u_i = 1$ is achieved if the surrogate server uploads only content ($\xi = infinity$). On the contrary, the value 0 is achieved if the surrogate server downloads only content. In the case of equal upload and download, the resulting value is 0.5.

Initially all the requested objects are stored in the origin server and all the CDN surrogate servers are empty. Taking into account the users requests, for each requested object, we find which is the best surrogate server in order to place it. In our context, the best surrogate server is the one that produces the maximum CDN utility value.

Then, we select from all the pairs of requested objectsurrogate server that have been occurred in the previous step, the one which produces the largest CDN utility value, and thus place this object to that surrogate server. The intuition of this placement is to maximize the utility of CDN surrogate servers. The above process is iterated until all the surrogate servers become full. As a result, a requested object may be assigned to several surrogate servers, but a surrogate server will have at maximum one copy of a requested object. The detailed algorithm is described in pseudocode in Figure 1. From the above, it occurs that the *utility* approach is heavily dependent on the incoming traffic. Specifically, previous research [9] has shown that the number of CDN surrogate servers, network proximity, congestion and traffic load of servers impact the resulting utility in a CDN. Concerning the complexity of the proposed algorithm is polynomial, since each phase requires polynomial time. In order to by-pass this problem, we may use content clusters (e.g., Web page communities) [4].

CDN Pricing. The CDN utility replica placement algorithm would be beneficial for CDN pricing. Specifically, we integrate the notion of net utility in the pricing model that has been presented in [1]. In particular, the monetary cost of the Web content provider under a CDN infrastructure is determined by the following equation:

$$U_{CDN} = V(X) + \tau(N) \times X - C_o - P(u) \tag{5}$$

where U_{CDN} is the final cost of Web content provider under a CDN infrastructure, $V(X)$ is the benefit of the content provider by responding to the whole request volume X, $\tau(N)$ is the benefit per request from faster content delivery through a geographically distributed set of N CDN surrogate servers, C_o is cost of outsourcing content delivery, $P(u)$ is the usage-based pricing function, and u is the CDN utility. Consequently, maximizing the CDN utility, $P(u)$ is also maximized and according to equation 5 the final cost of Web content provider under a CDN infrastructure (U_{CDN}) is minimized.

4 Simulation Testbed

CDN providers are real-time applications and they are not used for research purposes. Therefore, for the evaluation purposes, it is crucial to have a simulation testbed for the CDN functionalities and the Internet topology. Furthermore, we need a collection of Web users traces which access a Web server content through a CDN, as well as, the topology of this Web server content (in order to identify the Web page communities). Although we can find several users traces on the Web, real traces from CDN providers are not available. Thus, we are faced to use artificial data. Regarding the Web content providers, we use three real data sets. The data sets come from an active popular news Web content provider (BBC). Also, we take into account that the lifetime of Web objects is short. The expired Web objects are removed from the cache of surrogate servers. Table 1 presents the data sets used in the experiments.

This work is in line with the simulation-based evaluation of utility as described in [10]. We have developed a full simulation environment, which includes the following:

- a system model simulating the CDN infrastructure,
- a network topology generator,
- a client request stream generator capturing the main characteristics of Web users behavior.

Table 1. Summary of simulations parameters

Web site	BBC
Web site size	$568MB$
Web site number of objects	17000
Number of requests	1000000
Mean interarrival time of the requests	$1sec$
Distribution of the interarrival time	exponential
Requests stream z	0.5
Link speed	$1Gbps$
Network topology backbone type	AS
Number of routers in network backbone	3037
Number of surrogate servers	100
Number of client groups	100
Number of content providers	1
Cache size percentage of the Web site's size	6.25%, 12.5%, 25%, 50%

4.1 CDN Model

To evaluate the proposed approach, we used our complete simulation environment, called CDNsim [11], which simulates a main CDN infrastructure and is implemented in the C programming language. A demo can be found at http://oswinds.csd.auth.gr/~cdnsim/. It is based on the OMNeT++ library[1] which provides a discrete event simulation environment. All CDN networking issues, like CDN servers selection, propagation, queueing, bottlenecks and processing delays are computed dynamically via CDNsim, which provides a detailed implementation of the TCP/IP protocol (and HTTP), implementing packet switching, packet re-transmission upon misses, objects' freshness etc.

By default, CDNsim simulates a CDN with 100 CDN servers which have been located all over the world. Each CDN server in CDNsim is configured to support 1000 simultaneous connections. The default size of each surrogate server has been defined as the percentage of the total bytes of the Web server content. We also consider that each surrogate server cache is updated using a standard LRU cache replacement policy.

4.2 Network Topology

In a CDN topology we may identify the following network elements: CDN surrogate servers, origin server (Content Provider's main server), routers and clients. The routers form the network backbone where the rest of the network elements are attached. The distribution of servers and clients in the network affects the performance of the CDN. Different network backbone types result in different "neighborhoods" of the network elements. Therefore, the redirection of the requests and ultimately the distribution of the content is affected. In our testbed we use the AS Internet topology. The routers retransmit network packets using the TCP/IP protocol between the clients and the CDN. All the network phenomena such as bottlenecks and network delays, and packet routing protocols are simulated. Note that the AS Internet topology with a total of 3037 nodes captures a realistic Internet topology by using BGP routing data collected from a set of 7 geographically-dispersed BGP peers. In order to minimize the side effects due to intense network traffic we assume a high performance network with 1 Gbps link speed.

[1] http://www.omnetpp.org/article.php?story=20080208111358100

In order to ground our model in reality we have parameterized our CDN infrastructure with the real properties of a commercial CDN provider (Limelight). According to a recent study by [2][2], Limelight has clusters of servers deployed at 19 different locations around the world and each cluster has a different number of servers. In the United States there are 10 clusters, whereas, in Europe and Asia are seven clusters in total and Australia has only one. The rest of the world does not contain any cluster.

4.3 Requests Generation

As far as the requests stream generation is concerned, we used a generator, which reflects quite well the real users access patterns. Specifically, this generator, given a Web site graph, generates transactions as sequences of page traversals (random walks) upon the site graph, by modelling the Zipfian distribution to pages. In this work, we have generated 1 million users' requests. Each request is for a single object, unless it contains "embedded" objects. According to Zipfs law, the higher the value of z is the smaller portion of objects covers the majority of the requests. For instance, if $z = 0$ then all the objects have equal probability to be requested. If $z = 1$ then the probability of the objects fade exponentially. In this work we used the range 0.5 for z in order to capture an average case.

Then, the Web users' requests are assigned to CDN surrogate servers taking into account the network proximity and the surrogate servers' load, which is the typical way followed by CDN providers. Specifically, the following CDN redirection policy takes place: 1) A client performs a request for an object. 2) The request is redirected transparently to the closest surrogate server A in terms of network topology distance. 3) The surrogate server uploads the object, if it is stored in its cache. Elsewhere, the request is redirected to the closest to A surrogate server B, that contains the object. Then, the surrogate server A downloads the object from B and places it in its cache. If the object cannot be served by the CDN, the surrogate server A downloads the object from the origin server directly. 4) Finally the object is uploaded to the client.

5 Evaluation

In order to evaluate the proposed algorithm, we examine also the following heuristics: a) *Lat-cdn*: The outsourced objects are placed to surrogate servers with respect to the total network latency, without taking into account the objects popularity. Specifically, each surrogate server stores the outsourced objects which produce the maximum latency [7]; b) *il2p*: the outsourced objects are placed to the surrogate servers with respect to the total network latency and the objects load [6].

We evaluate the performance of CDN under regular traffic. It should be noted that for all the experiments we have a warm-up phase for the surrogate servers

[2] The paper has been withdrawn by Microsoft. We only use information about server locations from this work.

caches. The purpose of the warm-up phase is to allow the surrogate servers caches to reach some level of stability and it is not evaluated. The measures used in the experiments are considered to be the most indicative ones for performance evaluation.

5.1 Evaluation Measures

CDN Utility: It is the mean of the individual net utilities of each surrogate server in a CDN. The net utility is the normalized ratio of uploaded bytes to downloaded bytes. Thus, the CDN utility ranges in [0..1]. Using the notion of CDN utility we express the traffic activity in the entire CDN. Values over 0.5 indicate that the CDN uploads more content than it downloads through cooperation with other surrogate servers or the origin server. In particular, for the uploaded bytes we consider the content uploaded to the clients and to the surrogate servers. Values lower than 0.5 are not expected in the CDN schemes. The value 0.5 is an extreme case where each request is an object that has not been served by CDN.

Hit Ratio: It is the ratio of requests that has been served, without cooperation with other surrogate servers or the origin server, to the total number of requests. It ranges in [0..1]. High values of hit ratio are desired since they lead to reduced response times and reduced cooperation. Usually the hit ratio is improved by increasing the cache size and it is affected by the cache replacement policy.

Mean Response Time: It is the mean of the serving times of the requests to the clients. This metric expresses the users experience by the use of CDN. Lower values indicate fast served content.

Finally, we use the *t-test* to assess the reliability of the experiments. T-test is a significance test that can measure results effectiveness. In particular, the t-test would provide us an evidence whether the observed mean response time is due to chance. When a simulation testbed is used for performance evaluation, it is critical to provide evidence that the observed difference in effectiveness is not due to chance. For this purpose, we conducted a series of experiments, making random permutation of users (keeping the rest of the parameters unchanged).

5.2 Evaluation Results

The results are reported in Figures 2(a), 2(b) and 2(c), where x-axis represents the different values of cache size as a percentage of the total size of distinct objects (total Web site size).

Figure 2(a) depicts the CDN utility evolution that can be achieved by the CDN utility replica placement algorithm for different cache sizes. A general observation is that the proposed algorithm achieves high CDN utility values for any cache size comparing with the competitive ones. Taking a deeper look at the results, we observe a peak in the performance of the CDN utility at 25% cache size. This means that the *utility* approach achieves low data redundancy. Giving more insight to the performance peak, we should identify what happens to the

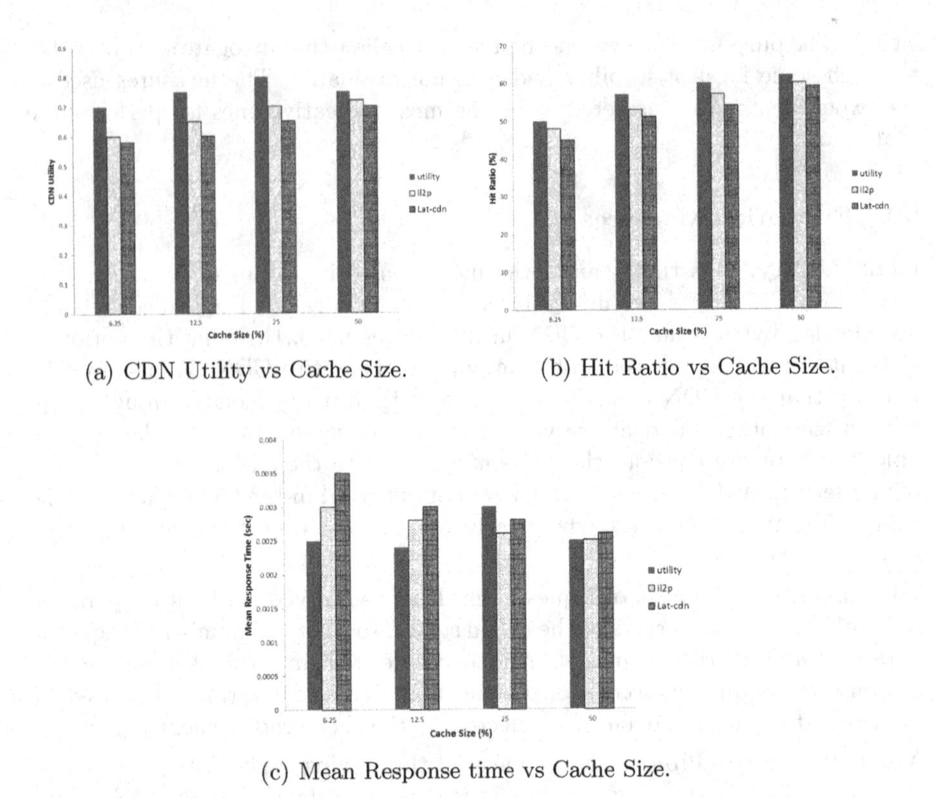

(a) CDN Utility vs Cache Size.

(b) Hit Ratio vs Cache Size.

(c) Mean Response time vs Cache Size.

Fig. 2. Evaluation results

eras before and after the peak. Before the peak, the cache size is quite small. Few replicas are outsourced to the CDN servers and the surrogate servers fail to cooperate. Most of the requests refer to objects that are not outsourced at all. Consequently, the surrogate servers refer to the origin server in order to gain copies of these objects. This leads to lower CDN utility as the surrogate servers upload less content. As the cache size increases, the amount of replicated content in the CDN increases as well. Therefore, the cooperation among the surrogate servers is now feasible and thus the CDN utility increases. After the peak, the cache size is large enough; consequently, this results in reducing the cooperation among the surrogate servers.

Figure 2(b) presents the hit ratio of the proposed algorithm for different values of cache sizes. As expected, the larger the cache size is, the larger the hit ratio is. Results show that the *utility* approach outperforms the other two examined policies. The most notable observation is that the proposed replication algorithm achieves high hit ratio for small cache sizes. Replicating a small size of Web site content, the observed performance peak guarantees satisfactory performance by reducing the traffic to origin server.

Figure 2(c) depicts the mean response time for different cache sizes. The general trend for the *utility* algorithm is that the response time is low for any cache size, observing for the *utility* approach a peak of mean response time at 25% cache size. The explanation for this peak is due to the fact that more requests are served; note that we also observe a peak in the performance of the CDN utility at 25% cache size (see Figure 2(a)).

Regarding the statistical test analysis, the t statistic is used to test the following null hypothesis (H_0):

H_0: *The observed mean response time and the expected mean response time are significantly different.*

We computed the $p - value$ of the test, which is the probability of getting a value of the test statistic as extreme as or more extreme than that observed by chance alone, if the null hypothesis H_0, is true. The smaller the $p - value$ is, the more convincing is the rejection of the null hypothesis. Specifically, we found that $p - value < 0,001$, which provides a strong evidence that the observed MRT is not due to chance.

6 Related Work

The growing interest in CDNs is motivated by a common problem across disciplines: how does one reduce the load on the origin server and the traffic on the Internet, and ultimately improve content delivery? In this direction, crucial data management issues should be addressed. A very important issue is the optimal placement of the requested content to CDN servers. Taking into account that this problem is NP complete, an heuristic method should be developed.

A number of research efforts have investigated the problem of content replication problem. Authors in [3] conclude that Greedy-Global heuristic algorithms are the best choice in making the replication. Replicating content across a geographically distributed set of servers and redirecting clients to the closest server in terms of latency has emerged as a common paradigm for improving client performance [5]. In [7], a self-tuning, parameterless algorithm, called lat-cdn, for optimally placing requested objects in CDNs surrogate servers is presented. Lat-cdn is based on network latency (an objects latency is defined as the delay between a request for a Web object and receiving that object in its entirety). The main advantage of this algorithm is that it does not require popularity statistics, since the use of them has often several drawbacks (e.g. quite a long time to collect reliable request statistics, the popularity of each object varies considerably etc.). However, this approach does not take into consideration the load of the objects (the load of an object is defined as the product of its access rate and size). Therefore, in this approach, it is possible to replicate in the same surrogate server objects with high loads and, thus, during a flash crowd event the server will be overloaded. To address this limitation, another approach is presented in [6] for optimally placing outsourced objects in CDN surrogate servers, integrating both the networks latency and the objects load.

However, all the existing approaches do not consider the CDN utility as a parameter to decide where to replicate the outsourced content. This is a key measure for CDNs since the notion of CDN utility can be used as a parameter to

CDN pricing policy. Typically, a CDN outsources content on behalf of content provider and charges according to a usage (traffic) based pricing function. In this context, several works exist which present the utility computing notion for content delivery [8,9,10]. They mostly provide description of architecture, system features and challenges related to the design and development of a utility computing platform for CDNs.

7 Conclusion

In this paper, we addressed the problem of optimally replicating the outsourced content in surrogate servers of a CDN. Under a CDN infrastructure (with a given set of surrogate servers) and a chosen content for delivery it is crucial to determine in which surrogate servers the outsourced content should be replicated. Differently from all other relevant heuristics approaches, we used the CDN utility metric in order to determine in which surrogate servers to place the outsourced objects. Summarizing the results, we made the following conclusions:

- The proposed algorithm achieves low replica redundancy. This is very important for CDN providers since low replica redundancy reduces the computing and network resources required for the content to remain updated. Furthermore, it reduces the bandwidth requirements for Web servers content, which is important economically for both individual Web servers content and for the CDN provider itself.
- The proposed algorithm achieves a performance peak in terms of CDN utility at a certain small cache size. Considering that the capacity allocation in surrogate servers affects the pricing of CDN providers we view this finding as particular important.

The experiment results are quite encouraging to further investigate the CDN utility replica placement algorithm under different traffic types and simulation parameters.

Acknowledgement. We are indebted to Theodoros Demetriou for his feedback on the early drafts of this work. This work was supported by the author's Startup Grant, funded by the University of Cyprus.

References

1. Hosanagar, K., Chuang, J., Krishnan, R., Smith, M.D.: Service adoption and pricing of content delivery network (cdn) services. Manage. Sci. 54, 1579–1593 (2008)
2. Huang, C., Wang, A., Li, J., Ross, K.W.: Measuring and evaluating large-scale cdns paper withdrawn at mirosoft's request. In: Proceedings of the 8th ACM SIGCOMM Conference on Internet Measurement, IMC 2008, pp. 15–29. ACM, New York (2008)
3. Kangasharju, J., Roberts, J., France Tlcom, R., Ross, K.W., Antipolis, S., Antipolis, S.: Object replication strategies in content distribution networks. Computer Communications, 367–383 (2001)

4. Katsaros, D., Pallis, G., Stamos, K., Vakali, A., Sidiropoulos, A., Manolopoulos, Y.: Cdns content outsourcing via generalized communities. IEEE Transactions on Knowledge and Data Engineering 21(1), 137–151 (2009)
5. Krishnan, R., Madhyastha, H.V., Srinivasan, S., Jain, S., Krishnamurthy, A., Anderson, T., Gao, J.: Moving beyond end-to-end path information to optimize cdn performance. In: Proceedings of the 9th ACM SIGCOMM Conference on Internet Measurement Conference, IMC 2009, pp. 190–201. ACM, New York (2009)
6. Pallis, G., Stamos, K., Vakali, A., Katsaros, D., Sidiropoulos, A., Manolopoulos, Y.: Replication based on objects load under a content distribution network. In: 22nd International Conference on Data Engineering Workshops, p. 53 (2006)
7. Pallis, G., Vakali, A., Stamos, K., Sidiropoulos, A., Katsaros, D., Manolopoulos, Y.: A latency-based object placement approach in content distribution networks. Web Congress, Latin American, 140–147 (2005)
8. Pathan, M., Broberg, J., Buyya, R.: Maximizing Utility for Content Delivery Clouds. In: Vossen, G., Long, D.D.E., Yu, J.X. (eds.) WISE 2009. LNCS, vol. 5802, pp. 13–28. Springer, Heidelberg (2009)
9. Pathan, M., Buyya, R.: A utility model for peering of multi-provider content delivery services. In: IEEE 34th Conference on Local Computer Networks, LCN 2009, pp. 475–482 (October 2009)
10. Stamos, K., Pallis, G., Vakali, A., Dikaiakos, M.D.: Evaluating the utility of content delivery networks. In: Proceedings of the 4th Edition of the UPGRADE-CN Workshop on Use of P2P, GRID and Agents for the Development of Content Networks, UPGRADE-CN 2009, pp. 11–20. ACM, New York (2009)
11. Stamos, K., Pallis, G., Vakali, A., Katsaros, D., Sidiropoulos, A., Manolopoulos, Y.: Cdnsim: A simulation tool for content distribution networks. ACM Trans. Model. Comput. Simul. 20, 10:1–10:40 (2010)
12. Tangpong, A., Kesidis, G.: A simple reputation model for bittorrent-like incentives. In: International Conference on Game Theory for Networks, GameNets 2009, pp. 603–610 (May 2009)
13. ul Islam, S., Stamos, K., Pierson, J.-M., Vakali, A.: Utilization-Aware Redirection Policy in CDN: A Case for Energy Conservation. In: Kranzlmüller, D., Toja, A.M. (eds.) ICT-GLOW 2011. LNCS, vol. 6868, pp. 180–187. Springer, Heidelberg (2011)

Using MINING@HOME for Distributed Ensemble Learning

Eugenio Cesario[1], Carlo Mastroianni[1], and Domenico Talia[1,2]

[1] ICAR-CNR, Italy
{cesario,mastroianni}@icar.cnr.it
[2] University of Calabria, Italy
talia@deis.unical.it

Abstract. MINING@HOME was recently designed as a distributed architecture for running data mining applications according to the "volunteer computing" paradigm. MINING@HOME already proved its efficiency and scalability when used for the discovery of frequent itemsets from a transactional database. However, it can also be adopted in several different scenarios, especially in those where the overall application can be divided into distinct jobs that may be executed in parallel, and input data can be reused, which naturally leads to the use of data cachers. This paper describes the architecture and implementation of the MINING@HOME system and evaluates its performance for the execution of ensemble learning applications. In this scenario, multiple learners are used to compute models from the same input data, so as to extract a final model with stronger statistical accuracy. Performance evaluation on a real network, reported in the paper, confirms the efficiency and scalability of the framework.

1 Introduction

The global information society is a restless producer and exchanger of huge volumes of data in various formats. It is increasingly difficult to analyze this data promptly, and extract from it the information that is useful for business and scientific applications. Fortunately, the notable advancements and the advent of new paradigms for distributed computing, such as Grids, P2P systems, and Cloud Computing, help us to cope with this data deluge in many scenarios.

Distributed solutions can be exploited for several reasons: (i) data links have larger bandwidths than before, enabling the assignment of tasks and the transmission of related input data in a distributed scenario; (ii) data caching techniques can help to reuse data needed by different tasks, (iii) Internet computing models such as the "public resource computing" or "volunteer computing" paradigm facilitate the use of spare CPU cycles of a large number of computers.

Volunteer computing has become a success story for many scientific applications, as a means for exploiting huge amount of low cost computational resources with a few manpower getting involved. Though this paradigm is clearly suited for the exploitation of decentralized architectures, the most popular volunteer computing platform available today, BOINC [2], assigns tasks according to a

A. Hameurlain et al. (Eds.): Globe 2012, LNCS 7450, pp. 100–111, 2012.

centralized strategy. Recent work, though, showed that the public computing paradigm can be efficiently combined with decentralized solutions to support the execution of costly data mining jobs that need to explore very large datasets. These reasons leaded to the design of the MINING@HOME architecture. In [11], simulation experiments showed that the architecture is able to solve the Closed Frequent Itemsets problem. After these early simulations that showed the benefits of the proposed approach, we worked to provide a full implementation of the framework and here we show that it is capable for the efficient execution of different data mining applications in a distributed scenario.

The main contribution of this work is the description of the MINING@HOME system architecture and implementation, and the evaluation of its performance when running "ensemble learning" applications in a real scenario. The ensemble learning approach combines multiple mining models together instead of using a single model in isolation [4]. In particular, the "bagging" strategy consists of sampling an input dataset multiple times, to introduce variability between the different models, and then extracting the combined model with a voting technique or a statistical analysis. MINING@HOME was profitably adopted to analyze a transactional dataset containing about 2 million transactions, for a total size of 350 MB. To run the application, the volunteer paradigm strategy is combined with a super-peer network topology that helps to exploit the multi-domain scenario adopted for the experiments.

The reminder of the paper is organized as follows: Section 2 presents the architecture, the involved protocols and the implementation of MINING@HOME . Section 3 discusses the ensemble learning strategy. Section 4 illustrates the scenario of the experiments and discusses the main results. Finally, Section 5 discusses related work and Section 6 concludes the paper.

2 Architecture and Implementation of Mining@home

As mentioned, the MINING@HOME framework was introduced in [11] to solve the problem of finding closed frequent itemsets in a transactional database. The system functionality and performance were only evaluated in a simulated environment. After that, MINING@HOME was fully implemented and was made able of coping with a number of different data analysis scenarios involving the execution of different data mining tasks in a distributed environment. The architecture of the MINING@HOME framework distinguishes between nodes accomplishing the mining task and nodes supporting data dissemination. In the first group:

- the **data source** is the node that stores the data set to be read and mined.
- the **job manager** is the node in charge of decomposing the overall data mining application in a set of independent tasks. This node produces a *job advert* document for each task, which describes its characteristics and specifies the portion of the data needed to complete the task. The job manager is also responsible for the collection of results.
- the **miners** are the nodes available for job execution. Assignment of jobs follows the "pull" approach, as required by the volunteer computing paradigm.

Data exchange and dissemination is done by exploiting the presence of a network of super-peers for the assignment and execution of jobs, and adopting caching strategies to improve the efficiency of data delivery. Specifically:

- **super peer** nodes constitute the backbone of the network. Miners connect directly to a super-peer, and super-peers are connected with one another through a high level P2P network.
- **data cachers** nodes operate as data agents for miners. In fact, data cachers retrieve input data from the data source or other data cachers, forward data to miners and store data locally to serve miners directly in the future.

The super-peer network allows the queries issued by miners to rapidly explore the network. The super-peer approach is chosen to let the system support several public computing applications concurrently, without requiring that each miner knows the location of the job manager and/or of the data cachers. Super-peers can also be used as rendezvous points that match job queries issued by miners with job adverts generated by the job manager.

The algorithm is explained here (see Figure 1). Firstly, the job manager partitions the data mining application in a set of tasks that can be executed in parallel. For each task, a "job advert" specifies the characteristics of the task to be executed and the related input data. An available miner issues a "job query" message to retrieve one of these job adverts. Job queries are delivered directly to the job manager, if it is possible. If the location of the latter is not known, job queries can travel the network through the super-peer interconnections (messages labeled with number 1 in the figure). When a job advert is found that matches the job query, the related job is assigned to the miner (message 2 in the figure). The miner is also informed, through the job advert, about the data that it needs to execute the job. The required input data can be the entire data set stored in the data source, or a subset of it.

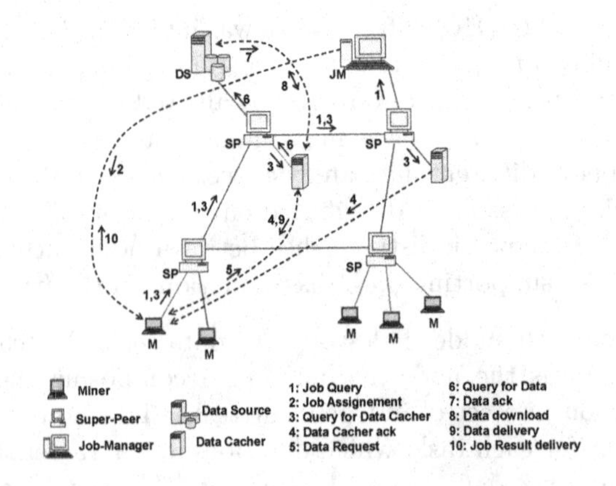

Fig. 1. Architecture of Mining@home

The miner does not download data directly from the data source, but issues a query to discover a data cacher (message 3). This query can find several data cachers, each of which sends an ack to the miner (message 4). After a short time interval, the miner selects the most convenient data cacher according to a given strategy (message 5), and delegates to it the responsibility of retrieving the required data. The data cacher issues a "data request" (message 6) to discover the data source or another data cacher that has already downloaded the needed data. The data cacher receives a number of acks from available data cachers (message 7), downloads the data from one of those (message 8), stores the data, and forwards it to the miner (message 9). Now the miner executes the task and, at its completion, sends the results to the job manager (message 10).

The algorithm can be used in the general case in which the location of job manager, data source and data cachers is unknown to miners and other data cachers. In ad hoc scenarios the algorithm can be simplified. Specifically, if the location of data source and data cachers are known, a job query (message 1) can be delivered directly to the job manager, instead of traveling the network, and messages 3-4 and 6-7 become unnecessary. Such simplifications are adopted for the experimental evaluation discussed in Section 4.

The MINING@HOME prototype has been implemented in Java, JDK 1.6. As depicted in Figure 1, the framework is built upon five types of nodes: job manager, data source, data cacher, super-peer and miner. Each node is multi-threaded, so that all tasks (send/receive messages, retrieve data, computation) are executed concurrently. Each miner exploits a *Mining Algorithm Library*, i.e., a code library containing the algorithms corresponding to the mining tasks.

3 Ensemble Learning and Bagging

Ensemble learning is a machine learning paradigm where multiple learners are trained to solve the same problem. In contrast to ordinary machine learning approaches, which try to learn one model from training data, ensemble methods build a set of models and combine them to obtain the final model. In a classification scenario, an ensemble method constructs a set of *base classifiers* from training data and performs classification by taking a vote on the predictions made by each classifier. As proved by mathematical analysis, ensemble classifiers tend to perform better (in terms of error rate) than any single classifier [12]. The basic idea is to build multiple classifiers from the original data and then aggregate their predictions when classifying unknown examples.

Bagging, also known as "bootstrap aggregating", is a popular ensemble learning technique [5]. Multiple training sets, or *bootstrap samples*, are sampled from the original dataset. The samples are used to train N different classifiers, and a test instance is labeled by the class that receives the highest number of votes by the classifiers. A logical view of the bagging method is shown in Figure 2. Each bootstrap sample has the same size as the original dataset. Since sampling is done with replacement, some instances may appear several times in the same bootstrap sample, while others may not be present. On average, a bootstrap

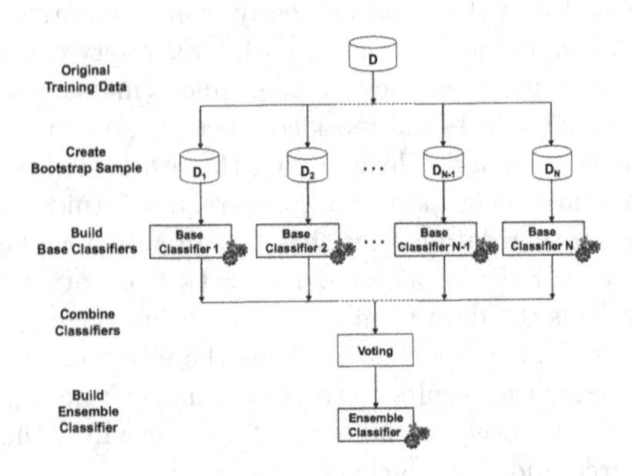

Fig. 2. A logical view of the bagging technique

sample D_i contains approximatively 63% of the original training data. In fact, if the original dataset contains n instances, the probability that a specific instance is sampled at least once is: $1 - (1 - 1/n)^n \rightarrow 1 - 1/e \simeq 0.631$, where the approximation is valid for large values of n. This implies that two different samples share, on average, about $0.631 \cdot 0.631 \approx 40\%$ of the n instances.

An application implementing the bagging technique can naturally exploit the MINING@HOME system. The described scenario matches the two main conditions that must hold in order to profitably exploit the features of MINING@HOME :

1. *Base Classifiers Are Independent.* Each base classifier can be mined independently from each other. Thus, it is possible to have a list of mining tasks to be executed, each one described by a distinct job descriptor. This fits the MINING@HOME architecture: each available miner is assigned the task of building one base classifier from a bootstrap sample of data, and at the end of execution the discovered classification model is transmitted to the job manager. Then, the miner may give its availability for a new job.
2. *Data Can Be Re-Used.* As mentioned before, in general different jobs need overlapping portions of input data, which is the rationale for the presence of distributed cache servers. After being assigned a job, the miner asks the input data to the closest data cacher, which may have already downloaded some of this data to serve previous requests. The data cacher retrieves only the missing data from the data source, and then sends the complete bootstrap sample to the miner. Of course, this leads to save network traffic and to a quicker response from the data cacher.

4 Experimental Evaluation

The performance of the MINING@HOME framework has been evaluated on a classification problem tackled with the bagging technique. We deployed the framework in a real network composed of two domains connected through a Wide Area

Network, as depicted in Figure 3. Each node runs an Intel Pentium 4 processor with CPU frequency 1.36GHz and 2GB RAM. The average inter-domain transfer rate is $197KB/s$, while the average intra-domain transfer rates are $918KB/s$ and $942KB/s$, respectively. The experiments were performed in a scenario where the job manager builds 32 base classifiers, by exploiting the bagging technique, on a transactional dataset D. The application proceeds as follows. The job manager builds a *job list*, containing the descriptions of the jobs that must be assigned to available miners. Each job is a request of building a J48 base classifier from a specific bootstrap sample. When all the jobs are executed, the job manager collects the extracted base classifiers and combines them to produce the final ensemble classifier.

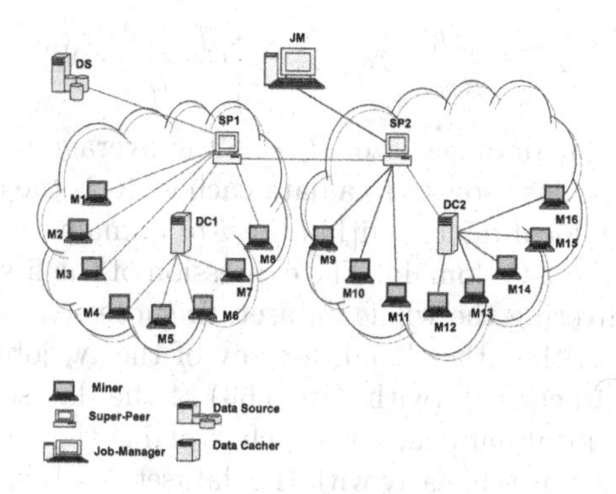

Fig. 3. Network architecture for the MINING@HOME experiments

The input dataset D is a subset of the *kddcup99* dataset[1]. The dataset, used for the KDD'99 Competition, contains a wide amount of data produced during seven weeks of monitoring in a military network environment subject to simulated intrusions. The input dataset used for the experiments is composed of 2 million transactions, for a total size of 350 MB.

Our evaluation followed two parallel avenues: we approximated performance trends analytically, for a generic scenario with N_D domains, and at the same time we compared experimental data to analytical predictions for the specific scenario described before. We adopted the techniques presented in [9] for the analysis of parallel and distributed algorithms. Let T_s be the *sequential execution time*, i.e., the time needed by a single miner to execute all the mining jobs sequentially, and T_o the *total overhead time* (mostly due to data transfers) experienced when the jobs are distributed among multiple miners. The total time spent by all the processors to complete the application can be expressed as:

$$nT_p = T_s + T_o \tag{1}$$

in which n is the number of miners and T_p is the parallel execution time when n miners are used in parallel. The speedup S - defined as the ratio between

[1] `http://kdd.ics.uci.edu/databases/kddcup99/kddcup99.html`

T_s and T_p - is then $S = \frac{T_s}{(T_s+T_o)/n} = \frac{nT_s}{T_s+T_o}$, and the efficiency E - defined as the ratio between the speedup and the number of miners - can be expressed as $E = \frac{1}{1+T_o/T_s}$. Therefore, the efficiency of the parallel computation is a function of the ratio between T_o and T_s: the lower this ratio, the higher the efficiency.

Let us start examining T_o. The total overhead time comprises the time needed to transfer data from the data source to data cachers and from these to miners. Both types of download are composed of a start up time (needed to open the connection and start the data transfer) and a time that is proportional to the amount of transferred data. Start up times are negligible with respect to the actual transfer time, therefore an approximation for T_o, in a scenario with N_D domains and N_{DC} data cachers, is:

$$T_o = \frac{|D|}{R_{DS}} \cdot N_{DC} + \sum_{i=1}^{N_D} (\frac{f \cdot |D|}{R_i} \cdot N_i) \quad (2)$$

where $|D|$ is the input data set size, R_{DS} is the average rate at which data is downloaded from the data source to a data cacher, R_i is the download rate from a data cacher to the local miners within the *i-th* domain, and N_i is the number of jobs assigned to the *i-th* domain. The expression of the first term derives from the necessity of delivering the whole dataset, in successive data transfers, to all the data cachers. On the other hand, for any of the N_i jobs that are executed under domain i, a fraction f (with $f \simeq 0.63$) of the dataset is downloaded by the miner from the local data cacher, which explains the second term.

Notice that T_o increases linearly with the dataset size $|D|$. On the other hand, the number of data cachers N_{DC} influences the two terms of expression (2) in different ways. The first term is proportional to N_{DC}, while the second term is inversely proportional to N_{DC} in an implicit fashion: when more data cachers are deployed on the network, possibly closer to miners, the intra-domain transfer rates R_i increase, and then the value of the second term decreases. Our experiments showed that the best choice is to have one data cacher per domain: this allows intra-domain transfer rates to be increased but at the same time avoids transmission of data to more data cachers than necessary, which would increase the first term in (2). This is the choice taken for our scenario, as shown in Figure 3. It is also possible to predict the impact of the number of miners. When two or more miners request data from the local data cacher at the same time, the intra-domain transfer rate R_i tends to decrease, because the uplink bandwidth of the data cacher is shared among multiple data connections. Since the probability of concurrent downloads increases when more miners are active in the same domain, the overall value of T_o tends to increase with the number of miners. This aspect will be examined in the comments to the experiments.

As opposed to T_o, the sequential time T_s does not depend on the network configuration, but of course it depends on the dataset size $|D|$. Specifically, the relationship between $|D|$ and T_s reflects the time complexity of the J48 algorithm, which is $O(|D| * ln|D|)$ [13].

Figure 4 reports the values of T_s and T_o measured in a scenario with two domains, one data cacher per domain, when varying the dataset size. The three

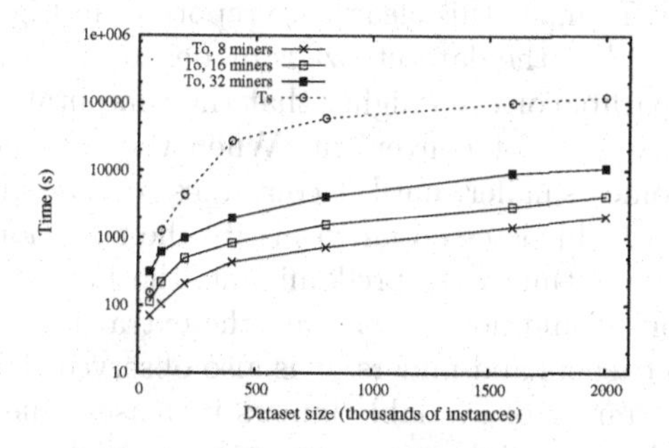

Fig. 4. Sequential time T_s and overhead time T_o vs. the size of the dataset, using 8, 16 and 32 miners

curves for T_o are obtained in the cases that 8, 16 and 32 miners are available to execute the 32 jobs. Of course, the values of T_s do not depend on the number of miners, as T_s is simply the sum of job computation times. The log scale is needed to visualize values that are very different from each other. The trends of T_s and T_o follow, quite closely, the theoretical prediction. This is shown in Figure 5, which compares the experimental values of T_s to the theoretical J48 complexity, and the experimental values of T_o, obtained with 16 miners, to those obtained with expression (2). While the gap between the two curves of T_o is small for any value of $|D|$, a larger discrepancy is observed for T_s when the dataset contains less than 500,000 instances, but this discrepancy tends to vanish for larger datasets. This is compatible with the fact that the expression $O(|D| * ln|D|)$ for the J48 theoretical complexity reflects an asymptotic behavior.

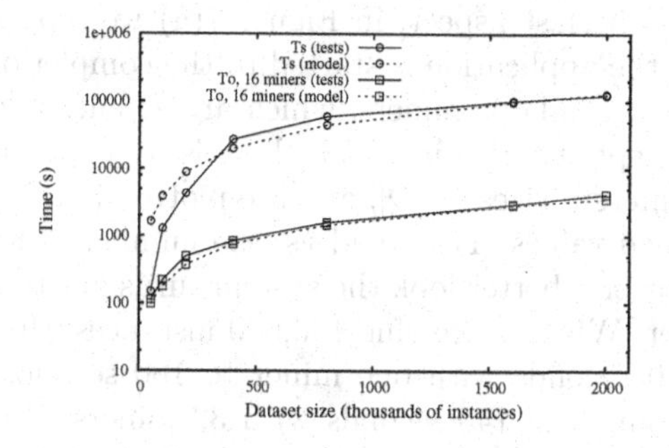

Fig. 5. Sequential time T_s and overhead time T_o (obtained with 16 miners) vs. the size of the dataset. Experimental results are compared to the theoretical prediction.

In Figure 4 it is also interesting to notice that the ratio T_o/T_s notably decreases in the first part the curve, which, as discussed before, is a sign that the efficiency of the architecture increases with the problem size. However, when the number of instances is higher than 500 thousands, the gap between T_s and T_o becomes stable, which in a graph with log scale means that the corresponding ratio becomes stable. To make this clearer we reported, in Figure 6, the relative weighs of T_s and T_o. When the dataset size is moderate, the time spent for data transmission is comparable or even higher that the computation time, therefore the distributed solution is not convenient. When the size increases, however, the weigh of T_o becomes smaller, until it converges to very small values, below 10% of the total time. This is essential to justify the adoption of a distributed solution: if the overhead time were predominant, the benefit derived from the parallelization of work would not compensate the extra time needed to transfer data to remote data cachers and miners. It is also observed that the weigh of T_o increases when the number of available miners increases, due to the impact of concurrent downloads of multiple miners from the same data cachers.

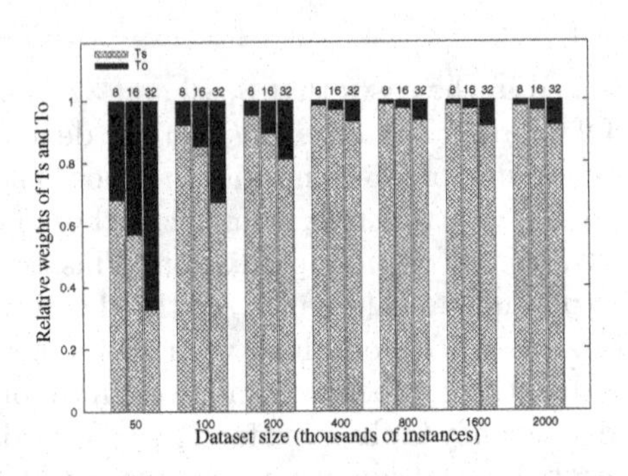

Fig. 6. Relative weights of sequential time T_s and overhead time T_o vs. the size of the dataset, using 8, 16 and 32 miners

To better analyze the last aspect, in Figure 7(a) we report the turnaround time (from the time the application is started to the completion time of the last job) vs. the number of available miners, which are equally partitioned between the two domains, except the case in which there is only one miner. Results are reported for three different sizes of $|D|$, and are plotted in a log scale due to the wide range of obtained values. The trend vs. the number of miners is the same at a first sight, but after a better look the system turns out to scale better when the problem is bigger. When processing 100,000 instances, the turnaround time decreases from 1,380 seconds with one miner to 160 seconds with 16 miners, but then nearly stabilizes to 145 seconds with 32 miners. With a dataset of 2 million instances, the turnaround time goes from about 34 hours with 1 miner to about 150 minutes with 16 miners and then continues to decrease to about 84 minutes with 32 miners. This indicates that using more and more miners can be efficient when the problem is big, but it is nearly useless – or even detrimental for the necessity of administrating a bigger system – when the problem size is limited. This is an index of good scalability properties, since scalable systems can be defined as those for which the number of workers that optimizes the performance increases with the problem size [9]. In Figure 7(b) we report the speedup, i.e., the ratio of the turnaround time obtained with a single node to

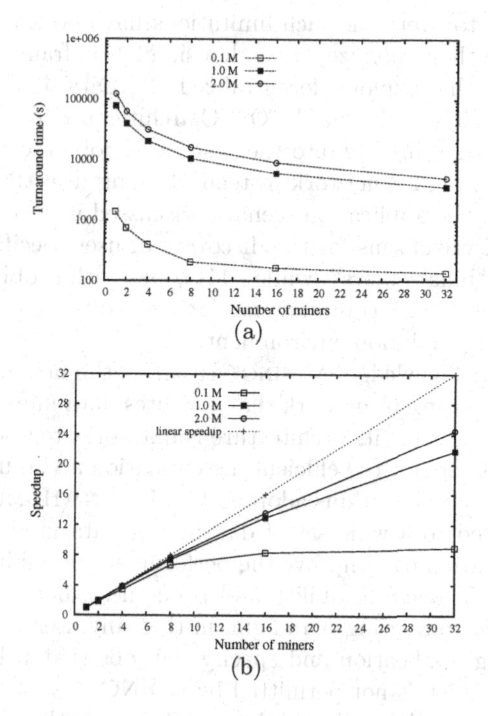

Fig. 7. Turnaround time (a) and speedup (b) vs. the number of available miners, for different values of the dataset size (1 M = 1 million instances)

the turnaround time computed with n nodes. It is clear that the speedup index saturates soon when the dataset is small, while the trend is closer to the optimal one (i.e., linear, shown for reference) as the dataset size increases.

5 Related Work

So far, the research areas of Distributed Data Mining and public resource computing have experienced little integration. The volunteer computing [1] paradigm has been exploited in several scientific applications (i.e., Seti@home, Folding@home, Einstein@home), but its adoption for mining applications is more challenging. The two most popular volunteer computing platforms available today, *BOINC* [2] and *XtremWeb* [6,8], are especially well suited for CPU-intensive applications but are somewhat inappropriate for data-intensive tasks, for two main reasons. First, the centralized nature of such systems requires all data to be served by a group of centrally maintained servers. Consequently, any server in charge of job assignment and data distribution is a clear bottleneck and a single point of failure for the system. Second, the client/server data distribution scheme does not offer valuable solutions for applications in which input data files can be initially stored in distributed locations or may be reused by different workers.

Some approaches to overcome such limitations have been recently proposed. In [7] and [11] the authors analyze, through a simulation framework, a volunteer computing approach that exploits decentralized P2P data sharing practices. The approach differs from the centralized BOINC architecture in that it seeks to integrate P2P networking directly into the system, as job descriptions and input data are provided to a P2P network instead of being directly delivered to the client. In particular, the application scenario discussed in [7] concerns the analysis of gravitational waveforms for the discovery of user specified patterns that may correspond to "binary stars", while [11] copes with problem of identifying closed frequent itemsets in a transactional dataset. So far, the analysis has only been performed in a simulation environment.

To the best of our knowledge, MINING@HOME is the first fully implemented public resource computing framework that executes data mining tasks over distributed data. In particular, its architecture is data-oriented because it exploits distributed cache servers for the efficient dissemination and reutilization of data files. The protocols and algorithms adopted by MINING@HOME are general and can be easily adapted to a wide set of distributed data mining problems. Not only this kind of solution can improve the performance of public computing systems, in terms of efficiency, flexibility and robustness, but also it can enlarge the use of the public computing paradigm, in that any user is allowed to define its own data mining application and specify the jobs that will be executed by remote volunteers, which is not permitted by BOINC.

More in general, several distributed data mining algorithms and systems have been proposed. In [3], a scalable and robust distributed algorithm for decision tree induction in distributed environments is presented. In order to achieve good scalability in a distributed environment, the proposed technique works in a completely asynchronous manner and offers low communication overhead. A distributed meta-learning technique is proposed in [10], where knowledge probing is used to extract descriptive knowledge from a black box model, such as a neural network. In particular, probing data is generated using various methods such as uniform voting, trained predictor, likelihood combination, etc. Differently from the classical meta-learning, the final classifier is learned from the probing data.

6 Conclusions

The public resource computing paradigm has proved useful to solve complex large problems in computational science areas, although it has not been used for data mining. In this paper we presented a software system, called MINING@HOME , which exploits that paradigm to implement large-scale data mining applications in a decentralized infrastructure. The developed system can be profitably used in the internet for mining massive amount of data available in remote Web sites or geographically dispersed data repositories. We evaluated the system and its performance on a specific use case, in which classification of instances is driven by the use of ensemble learning techniques. The system can be used for the execution of other mining applications, when these can be decomposed in smaller jobs and can exploit the presence of distributed data cachers. Further applications of MINING@HOME are currently under investigation.

Acknowledgments. This research work has been partially funded by the MIUR projects FRAME (PON01_02477) and TETRIS (PON01_00451).

References

1. Anderson, D.P.: Public computing: Reconnecting people to science. In: Proceedings of Conference on Shared Knowledge and the Web, Madrid, Spain, pp. 17–19 (2003)
2. Anderson, D.P.: Boinc: A system for public-resource computing and storage. In: GRID 2004: Proceedings of the Fifth IEEE/ACM International Workshop on Grid Computing (GRID 2004), Washington, DC, USA, pp. 4–10 (2004)
3. Bhaduri, K., Wolff, R., Giannella, C., Kargupta, H.: Distributed decision tree induction in peer-to-peer systems (2008)
4. Bishop, C.M.: Pattern Recognition and Machine Learning. Springer (2006)
5. Breiman, L.: Bagging predictors. Machine Learning 24(2), 123–140 (1996)
6. Cappello, F., Djilali, S., Fedak, G., Herault, T., Magniette, F., Neri, V., Lodygensky, O.: Computing on large-scale distributed systems: Xtrem web architecture, programming models, security, tests and convergence with grid. Future Generation Computer Systems 21(3), 417–437 (2005)
7. Cozza, P., Mastroianni, C., Talia, D., Taylor, I.: A Super-Peer Model for Multiple Job Submission on a Grid. In: Lehner, W., Meyer, N., Streit, A., Stewart, C. (eds.) Euro-Par Workshops 2006. LNCS, vol. 4375, pp. 116–125. Springer, Heidelberg (2007)
8. Fedak, G., Germain, C., Neri, V., Cappello, F.: Xtremweb: A generic global computing system. In: Proceedings of the IEEE Int. Symp. on Cluster Computing and the Grid, Brisbane, Australia (2001)
9. Grama, A.Y., Gupta, A., Kumar, V.: Isoefficiency: Measuring the scalability of parallel algorithms and architectures. IEEE Concurrency 1, 12–21 (1993)
10. Guo, Y., Sutiwaraphun, J.: Probing Knowledge in Distributed Data Mining. In: Zhong, N., Zhou, L. (eds.) PAKDD 1999. LNCS (LNAI), vol. 1574, pp. 443–452. Springer, Heidelberg (1999)
11. Lucchese, C., Mastroianni, C., Orlando, S., Talia, D.: Mining@home: Towards a public resource computing framework for distributed data mining. Concurrency and Computation: Practice and Experience 22(5), 658–682 (2010)
12. Tan, P.N., Steinbach, M., Kumar, V.: Introduction to Data Mining. Pearson International Edition (2006)
13. Witten, I.H., Frank, E.: Data mining: practical machine learning tools and techniques with Java implementations. Morgan Kaufmann (2000)

Author Index